Springer Theses

Recognizing Outstanding Ph.D. Research

Aims and Scope

The series "Springer Theses" brings together a selection of the very best Ph.D. theses from around the world and across the physical sciences. Nominated and endorsed by two recognized specialists, each published volume has been selected for its scientific excellence and the high impact of its contents for the pertinent field of research. For greater accessibility to non-specialists, the published versions include an extended introduction, as well as a foreword by the student's supervisor explaining the special relevance of the work for the field. As a whole, the series will provide a valuable resource both for newcomers to the research fields described, and for other scientists seeking detailed background information on special questions. Finally, it provides an accredited documentation of the valuable contributions made by today's younger generation of scientists.

Theses are accepted into the series by invited nomination only and must fulfill all of the following criteria

- They must be written in good English.
- The topic should fall within the confines of Chemistry, Physics, Earth Sciences, Engineering and related interdisciplinary fields such as Materials, Nanoscience, Chemical Engineering, Complex Systems and Biophysics.
- The work reported in the thesis must represent a significant scientific advance.
- If the thesis includes previously published material, permission to reproduce this must be gained from the respective copyright holder.
- They must have been examined and passed during the 12 months prior to nomination.
- Each thesis should include a foreword by the supervisor outlining the significance of its content.
- The theses should have a clearly defined structure including an introduction accessible to scientists not expert in that particular field.

More information about this series at http://www.springer.com/series/8790

Johanna Gramling

Search for Dark Matter with the ATLAS Detector

Probing Final States of Missing Energy and an Energetic Jet or Top Quarks

Doctoral Thesis accepted by
the University of Geneva, Switzerland

Author
Dr. Johanna Gramling
Department of Physics and Astronomy
University of California
Irvine, CA, USA

Supervisor
Prof. Dr. Xin Wu
University of Geneva
Geneva, Switzerland

ISSN 2190-5053   ISSN 2190-5061 (electronic)
Springer Theses
ISBN 978-3-319-95015-0   ISBN 978-3-319-95016-7 (eBook)
https://doi.org/10.1007/978-3-319-95016-7

Library of Congress Control Number: 2018946679

© Springer International Publishing AG, part of Springer Nature 2018
This work is subject to copyright. All rights are reserved by the Publisher, whether the whole or part of the material is concerned, specifically the rights of translation, reprinting, reuse of illustrations, recitation, broadcasting, reproduction on microfilms or in any other physical way, and transmission or information storage and retrieval, electronic adaptation, computer software, or by similar or dissimilar methodology now known or hereafter developed.
The use of general descriptive names, registered names, trademarks, service marks, etc. in this publication does not imply, even in the absence of a specific statement, that such names are exempt from the relevant protective laws and regulations and therefore free for general use.
The publisher, the authors, and the editors are safe to assume that the advice and information in this book are believed to be true and accurate at the date of publication. Neither the publisher nor the authors or the editors give a warranty, express or implied, with respect to the material contained herein or for any errors or omissions that may have been made. The publisher remains neutral with regard to jurisdictional claims in published maps and institutional affiliations.

Printed on acid-free paper

This Springer imprint is published by the registered company Springer International Publishing AG part of Springer Nature
The registered company address is: Gewerbestrasse 11, 6330 Cham, Switzerland

Für Milan und für Lionel.

Supervisor's Foreword

The evidence that some invisible ("dark") matter exists in the universe was first observed many decades ago from studying the relative movement of stars and galaxies, and was more recently confirmed by gravitational lensing effects. It is estimated that about 27% of total mass and energy in the universe is in this form of Dark Matter (DM). In spite of decades of search effort, the nature of the DM remains a mystery and has become one of the hottest research subjects in fundamental physics. The research work of Johanna Gramling reported in this thesis represents a significant contribution to this quest.

With the latest progress in cosmology, in particular, the precise measurement of the anisotropy of the Cosmic Microwave Background (CMB), the discovery of the Dark Energy, and the study of the large-scale structure of the Universe, a model that describes reasonably well the observations, the so-called "Standard Model of Big Bang Cosmology" (ΛCDM) has emerged, in which DM makes up about 27% of the mass–energy density of the universe. One of the possible candidates of the DM is a new type of particle.

Dark Matter particles can be searched for by observing their interactions with a detector as the Earth moves through the galactic DM halo, or by the detection of secondary particles produced from DM annihilation or decay. These two types of searches are referred to as "direct" and "indirect" searches. The search described in this thesis belongs to the third category, the so-called "collider search", since DM particles can also be produced at particle colliders.

Johanna Gramling's thesis describes a DM search using data collected by the ATLAS experiment at the current most powerful particle accelerator in the world, the Large Hadron Collider (LHC) at CERN. One refreshing aspect of this thesis is that it goes far beyond the usual technical description of a data analysis, typical for a thesis in particle physics. It also contains an in-depth analysis of the theoretical framework and the methodology related to the Effective Field Theory (EFT) and simplified models, which are needed to interpret the search results as model independent as possible, since little is known about the DM production mechanism.

The subject of the EFT is introduced in Chap. 2 and fully developed in Chap. 5. EFT is a useful approach to characterize new physics without having to build complex models, which has the advantage that the dependency on the details of the models is reduced, thus the same experimental search can be applied to a large set of similar models. However, for this approach to be valid, conditions related to the scale of the new physics and the experimental condition need to be met. The result obtained by this research indicated indeed that the EFT validity is limited in many cases, and a rescaling of the obtained DM limits is necessary. In addition, the EFT approach should be complemented by the simplified model approach that avoids the validity problem, but with somewhat increased model dependency. Following this work, this scaling procedure, as well as the complementary EFT/simplified model approach, has been adopted by the LHC experiments.

Two DM searches are reported in this thesis. One is with the "mono-jet" final state, and another using the final states with "missing energy and top quarks in the one-lepton channel." The technical details of the data analysis are clearly explained, and in relation to the methodology studies introduced above, the interpretations of the search results are comprehensive and new limits are derived within the new framework of simplified models.

The search for DM in the mono-jet final states, described in Chap. 6, is done with the data sample collected by ATLAS during 2012 when LHC was colliding protons at a center-of-mass energy of 8 TeV. Mono-jet is one of the most inclusive DM search channels in which the recoiling particle of produced the DM particle, which escaped detection due to its very weak interaction with matter, is a jet. No DM signal is observed, so the result is used to constrain the DM production parameters using both the EFT and simplified model. In addition, the mono-jet search, together with two other ATLAS DM searches, namely, the mono-Z with leptonic Z decay search and the mono-W/Z with hadronic decays search, were used to derive limits on three well-motivated simplified models: a Z' model, an s-channel axial-vector mediator model, and a t-channel scalar mediator model. The study, reported in Chap. 7, showed that using simplified models allowed to compare and optimize DM searches in different channels globally in a consistent way.

The search for DM using the final states with missing energy and top quarks in the one-lepton channel, described in Chap. 8, is related to the Supersymmetry (SUSY), introduced in Chap. 3. An important aspect of SUSY related to this thesis is that it introduces new particles, and in one category of the models (R-parity conserved), one of the new particles, the neutralino, is an excellent candidate for Dark Matter. The search is focused on the supersymmetry partner of the top quark, the stop, using the final states with missing energy and top quarks in the one-lepton channel. The same search is also sensitive to DM production in association with top quarks. Several signal regions are used to optimally search for the new particle in different mass ranges. Overall, there are good agreements between observations and background expectations. The limits on the DM are derived for simplified models with scalar or pseudo-scalar mediators and presented as upper bound on the combined coupling, and as exclusion regions in the plane of DM mass versus mediator mass.

Besides the outstanding quality of its research results, the thesis of Johanna Gramling is also an excellent reference for particle physicists interested in understanding the latest methodology used to search for new phenomenon in particle physics, in particular, those related to the search of DM at particle colliders. With the planned upgrades and the long-term running of the LHC, DM searches represented by this thesis will continue to be one of the research highlights of LHC experiments in the next decade.

Geneva, Switzerland
March 2018

Prof. Xin Wu

Abstract

The overwhelming astrophysical evidence for Dark Matter is an important motivation to search for new physics at the Large Hadron Collider (LHC) at CERN. While the Standard Model of particle physics is able to predict measurements and observations to an astounding precision, it does not provide a candidate particle for Dark Matter. If possibly produced in high-energy proton–proton collisions, such particles would traverse the detectors without leaving a signal. Hence, searches rely on the resulting momentum imbalance in the transverse plane. One particular extension of the Standard Model that allows for a Dark Matter candidate is supersymmetry. Since the supersymmetric partner of the top quark is expected to be relatively light, it could be in reach of LHC experiments and possibly detected.

This thesis presents a study of the validity of commonly used effective field theory models of Dark Matter production at the LHC. It shows that in a significant fraction of events the assumptions of an effective field theory description are not justified, which requires a redefinition of strategy when interpreting LHC results in terms of Dark Matter production. The results from a search for new phenomena in events with an energetic jet and large missing transverse energy are presented. It is performed on 20.3 fb^{-1} of 8 TeV pp collision data, recorded by the ATLAS detector at the LHC. No evidence for new physics was observed. The results are interpreted in terms of Dark Matter production within an effective model as well as using a simplified model, motivated by the findings of the validity study. Subsequently, this result and two other ATLAS searches are studied in a detailed reinterpretation in terms of simplified models of Dark Matter production. A large range of parameters is tested for three different simplified models. The study revealed that a dedicated optimization in view of simplified models would be beneficial, especially in the regime of small missing transverse energy. Final states of Dark Matter and top quarks are well motivated by models with a scalar or pseudo-scalar particle mediating the interaction between Standard Model and Dark Matter particles. The resulting final state is similar to that of the production of supersymmetric top partners. A search for new phenomena in such final states of top quark pairs and large missing transverse energy, performed on 13.2 fb^{-1} of data from 13 TeV *pp* collisions, is presented. An excess of data events over the Standard Model background prediction of 3.3σ was observed in a signal region optimized for Dark Matter signals. Interpretations of the results are presented for two decay scenarios of supersymmetric top quark partners and for Dark Matter production in association with top quarks.

Acknowledgements

The completion of this thesis marks the end of my Ph.D. period. I had the opportunity to work on interesting physics questions, meet interesting people, make interesting experiences, some better than other, and of course, learn a lot. However, this would all not have been possible without the support and help of many people, colleagues, and friends.

I want to thank my Profs. Xin Wu and Giuseppe Iacobucci for their support and guidance through the last years. Also Prof. Allan Clark provided very helpful advise and encouragement. Lucian deserves a very special thanks for his support in matters of physics and politics: he was never tired of passing by my office, encouraging me, and standing up for me when necessary. I also want to thank all current and former members of the Geneva ATLAS group for their support and company. Thank you to Steven and Luis who did their best to solve computing emergencies that of course always occur in the worst moments. Steven, thank you for always answering my questions about monojets, Dark Matter, jets, and recently the jet trigger.

I would like to thank the people directly involved in improving this thesis in several ways, especially the committee members Andreas Hoecker, Toni Riotto, and Tobias Golling for their extremely helpful and interesting comments. Thanks to Andrea (and Mike), Stefan, Lucian, Steven, Priscilla, Kristof, Philippe, Michelle, and Jan for reading and commenting my drafts–your input was incredibly valuable and allowed me to improve the quality of my thesis significantly.

Furthermore, I would like to thank all the people I had the pleasure of working with in the various analyses and projects during my Ph.D.: the dark monojets Ruth, Steven and Philippe, the stop 1L team, especially Till for his seemingly endless enthusiasm, Priscilla for many insightful Dark Matter discussions, Jan and Daniela for continuous SWup support, and Sophio, Javier, and Keisuke for working through so many nights with me before ICHEP. For FTK support, I want to thank Alberto, Guido, Karol, and Takashi for answering my numerous questions and guiding me from having no clue about hardware to being able to run and debug AMB tests.

I also want to thank Hajo, Maida and Aleksi, and Alix to jump in to spent time with Lionel when it was necessary. Thanks to Nils and Kristof for meeting up with

Herr Schwitter–some days the best motivation to come to work. Thank you Oliv', Jaelle, Catherine, and Sabine for forcing me to stop thinking about work for a few hours. I would like to warmly thank Marie and Russ, Lisa, Marie, and Tuva for being there, listening to my complaints, understanding the long silence from my side, and being happy with me that it is finally done.

I am very thankful to my family, my parents, and my brother, for all their support and effort to understand what I am doing and why. Even though it was not always easy to bring our worlds together in this dense period, I know you were thinking of me a lot. Thank you Lionel, for making me smile, even in dark moments.

Most importantly, I want to thank Jan, for standing next to me, holding me, keeping me sane and safe, taking as much off my shoulders as possible, supporting me in all possible ways, for everything, beyond words, for being such a very special person.

Contents

1	**Introduction**	1
	References	3
2	**The Standard Model**	5
	2.1 Symmetries	7
	2.2 Strong Interactions	8
	2.3 Electroweak Interactions	9
	2.3.1 Spontaneous Symmetry Breaking	10
	2.3.2 Fermion Masses	11
	2.4 Defining the Standard Model	14
	2.4.1 Parameters	15
	2.5 Open Questions and Known Problems of the Standard Model	15
	2.5.1 Neutrino Masses	15
	2.5.2 Fine Tuning and Naturalness	16
	2.5.3 Further Questions	18
	References	19
3	**Dark Matter**	21
	3.1 Evidence for Dark Matter	21
	3.1.1 Galactic Scales: Galaxy Rotation Curves	22
	3.1.2 Cluster Scales: Gravitational Lensing	23
	3.1.3 Cosmological Scales: Cosmic Microwave Background	23
	3.2 Properties of Dark Matter	25
	3.3 Thermal Dark Matter	26
	3.4 Candidates for Non-baryonic Dark Matter	28

	3.5	Searches for Dark Matter	29
		3.5.1 Direct Detection	29
		3.5.2 Indirect Detection	31
		3.5.3 Dark Matter at the LHC	32
	3.6	Effective Field Theory Description of Dark Matter	33
		3.6.1 Interplay Between Dark Matter Searches	36
	References		37
4	**Supersymmetry**		**41**
	4.1	Main Concepts of Supersymmetry	42
	4.2	The Minimal Supersymmetric Standard Model	43
		4.2.1 Breaking of Supersymmetry	43
		4.2.2 R-Parity	45
		4.2.3 Reducing Parameters	45
	4.3	Supersymmetric Top Quark Partners	47
	References		50
5	**The ATLAS Detector at the LHC**		**51**
	5.1	Particle Collisions at the LHC	52
	5.2	The ATLAS Detector	56
		5.2.1 Coordinate System and Variable Definitions	56
		5.2.2 Detector Design	57
	5.3	Data Acquisition and Trigger	61
		5.3.1 Missing Transverse Energy Trigger	62
		5.3.2 The Fast TracKer (FTK)	63
	5.4	Physics Object Definitions	68
		5.4.1 Track and Vertex Reconstruction	68
		5.4.2 Muons	69
		5.4.3 Electrons and Photons	69
		5.4.4 Jets	70
		5.4.5 B-Jets	71
		5.4.6 Hadronically Decaying Tau Leptons	72
		5.4.7 Missing Transverse Energy	73
	5.5	Event Simulation	73
		5.5.1 Sample Generation	73
		5.5.2 Detector Simulation	75
	References		76
6	**Validity of Effective Field Theory Dark Matter Models at the LHC**		**81**
	6.1	Effective Field Theory Models of Dark Matter and Their Validity	82
	6.2	Analytical Analysis of the Effective Field Theory Validity	83
		6.2.1 Quantifying the Effective Field Theory Validity	85

	6.3	Numerical Approach to Effective Field Theory Validity	89
		6.3.1 Simulation and Analysis Description	89
		6.3.2 Results	90
	6.4	Implications on Dark Matter Searches at LHC	92
	6.5	Conclusions	95
	References		96
7	**Search for Dark Matter in Monojet-like Events**		**97**
	7.1	Analysis Strategy	98
	7.2	Dataset and Simulations	99
		7.2.1 Dataset	99
		7.2.2 Monte Carlo Simulations	99
	7.3	Event Selection	102
		7.3.1 Reconstructed Objects	102
		7.3.2 Preselection	103
		7.3.3 Veto on Isolated Tracks	103
		7.3.4 Cut Optimisation for Dark Matter Signals	114
		7.3.5 Signal Region Definition	116
	7.4	Background Estimation	116
		7.4.1 W/Z+jets Background	118
		7.4.2 Multijet Background	122
		7.4.3 Non-collision Background	123
	7.5	Systematic Uncertainties	124
		7.5.1 Uncertainties on the Background Prediction	124
		7.5.2 Signal Systematic Uncertainties	125
	7.6	Results	127
	7.7	Interpretation	127
		7.7.1 Model-Independent Limits	127
		7.7.2 Dark Matter Pair Production	128
	7.8	Conclusions	134
	References		137
8	**Constraints on Simplified Dark Matter Models from Mono-X Searches**		**141**
	8.1	Simplified Models of Dark Matter Production	142
		8.1.1 Mono-X Signatures	143
		8.1.2 Mass and Coupling Points	144
		8.1.3 Mediator Width	145
		8.1.4 Rescaling Procedure	146
	8.2	Recasting Mono-X Constraints	147
		8.2.1 Signal Simulation	148
		8.2.2 Limit Setting Strategy	149
		8.2.3 Monojet Channel	150

		8.2.4	Mono-Z(lep) Channel	152
		8.2.5	Mono-W/Z(had) Channel	154
	8.3	Results and Discussion	155
		8.3.1	Limits on the Couplings	155
		8.3.2	Comparison with Relic Density Constraints	162
		8.3.3	Comparison with Direct Detection Constraints	163
	8.4	Conclusions	164
	References	165
9	Search for New Physics in Events with Missing Energy and Top Quarks	169
	9.1	Analysis Strategy	171
	9.2	Dataset	171
		9.2.1	Monte Carlo Simulations	172
		9.2.2	Trigger Selection	174
	9.3	Event Reconstruction and Selection	177
		9.3.1	Object Definition	177
	9.4	Discriminating Variables	180
	9.5	Signal Regions	182
		9.5.1	Preselection	182
		9.5.2	Dark Matter Optimisation	183
		9.5.3	Signal Region Overview	190
	9.6	Background Estimation	190
		9.6.1	Control Regions	192
		9.6.2	Validation Regions	195
	9.7	Systematic Uncertainties	200
		9.7.1	Experimental Uncertainties	201
		9.7.2	Theoretical Uncertainties	201
	9.8	Results	202
	9.9	Interpretation of the Results	205
		9.9.1	Limits on Dark Matter Models	206
		9.9.2	Limits on Direct Stop Production	207
	9.10	Conclusions	210
	References	211
10	**Conclusions**	215

Appendix A: Additional Aspects of Dark Matter and Its Properties 219

Appendix B: Differential Cross-Sections for Additional Effective Operators 227

Appendix C: Auxiliary Material for the *Stop* Analysis 229

Appendix D: Investigation of the Data Excess 275

Chapter 1
Introduction

Everything we know today about the world around us we owe to human curiosity, which prevents us from settling down with an answer, and makes us investigate further and ask for deeper reasons. Already Johann Wolfgang Goethe's Faust wanted to know "what keeps the world together at heart" ("was die Welt im Innersten zusammenhält")–and he failed [1]. In the end he had to realise that science–and he did not restrict himself to natural sciences alone–is not able to provide any satisfying answer. On a similar note, Immanuel Kant concerned himself, in his antinomies of pure reason [2], with questions like: is there a smallest, elementary building block of matter? Does the world have a beginning and an end in space and time? Is everything in the world fully determined? Does something exist that is not part of our world? In a nutshell, Kant introduces a distinction between, on the one hand, the total of what we can possibly know and experience, and, on the other hand, how the world "truly" intrinsically is. Since the above questions concern the intrinsic properties of nature, we cannot even hope to obtain any answers. At the same time, Kant states that we cannot stop asking these questions, even though we know that we will never be able to answer them–it is part of what makes us human.

The above questions are asked in similar ways also physics and are indeed what ultimately motivates fundamental physics to continue further. There are very successful and precise models describing many observations–but these are models, descriptions, often with clear limitations, and we cannot say if there is any correspondence to how the world "truly" is. Some argue that the simplicity, the (mathematical) beauty the models might achieve is a sign that they have to be at least close to the truth. But, strictly speaking, we cannot be sure we learn anything about the "true" world apart from obtaining a very accurate description, and reliable measurement predictions.

But why pursue science, physics at all? Apart from all practical reasons, I would say, even if we are not sure, it seems like the best option to at least approach answers to these fundamental questions. Physics can be considered as the most fundamental natural science, it spans from the smallest to the largest possible scales. Particle physics, in my opinion, has a unique role: not only is it directly concerned with the smallest possible scales and the fundamental building blocks of matter, but through the history of the universe it is tightly connected to the largest possible scales and cosmology.

One example for this interplay is *Dark Matter*, established to explain astrophysical observations: Dark Matter does not interact with ordinary matter, except by the very weak gravitational force and maybe an additional, weak interaction–it is a bit like a parallel world. On the other hand, without Dark Matter we would not exist: without Dark Matter, galaxies, galaxy clusters and eventually stars like the sun would not have been able to coalesce, heavy elements, would not have been created, and so on. So Dark Matter had a crucial role in the very early history of the universe, earth, life and us. The impact on cosmology is evident: Dark Matter, making up more than 84% of the matter in the universe, can alter cosmological predictions significantly, for example through its influence on large-scale structure formation. But Dark Matter is also studied in particle physics: if it is made up of particles–a plausible hypothesis– these particles would have to be of a new, unknown kind, since none of the established particles is an eligible candidate. Proposed candidates do not only have to respect the bounds of particle physics experiments searching for Dark Matter but also have to be consistent with cosmological constraints.

The study of particle collisions emerged as a powerful way to probe the interactions of fundamental particles. If some weak interaction is assumed between Dark Matter and ordinary matter apart from gravity, colliders present an interesting possibility to search for Dark Matter, since it might be produced in the collisions. However, due to their very weak interaction with the detector material such particles would not leave a signal in the detector. Searching for their production is hence challenging: the analyses rely on missing momentum in the transverse plane, caused by the invisible particles recoiling against visible objects. The Large Hadron Collider (LHC) at CERN provides proton–proton collisions at unprecedented centre-of-mass energies and offers a unique opportunity to search for new physics. The ATLAS experiment records the particle collisions and allows to evaluate these events, for example in view of Dark Matter searches.

This thesis discusses several aspects of Dark Matter searches at the LHC and their interpretation. In Chaps. 2–5, the Standard Model of particle physics and its open questions are introduced and an overview of Dark Matter properties and experimental searches is presented, before the main concepts of Supersymmetry are outlined and the ATLAS detector is explained. Chap. 6 discusses the interpretation of Dark Matter searches at the LHC in terms of commonly used effective field theory models. The presented study revealed that the assumptions of an effective approach are widely violated in collisions at LHC energies.

Subsequently, the search for new physics in final states of large missing transverse energy and an energetic jet, using 20.3 fb^{-1} of pp collision data at 8 TeV, is described in Chap. 7. Such final states can originate from the pair production of Dark Matter, recoiling against a jet. This is a general scenario of Dark Matter production at a hadron collider, and commonly the most sensitive channel. The results are interpreted in terms of Dark Matter production within an effective model as well as using a

Simplified Model, considering the findings of the validity study. In Chap. 8, this and two other ATLAS searches for Dark Matter are reinterpreted in terms of Simplified Models of Dark Matter production, considering a broad parameter range. The models assume one mediating particle in addition to the Dark Matter particles and try to circumvent the validity issues of effective approaches while resting general.

The Standard Model of particle physics cannot provide a candidate for Dark Matter. Furthermore, despite its success, it exhibits problems and leaves important questions unanswered. One proposed extension, Supersymmetry, is especially attractive since it is able to offer solutions to several problems at once. Furthermore, it can offer a candidate for Dark Matter. Since the supersymmetric partner of the top quark is expected to be lighter than other new particles, searching for these top squarks at the LHC is a promising approach. Such signals lead to similar final states as Dark Matter production in association with top quarks, well motivated in the case of (pseudo-)scalar mediators. A search for new phenomena in final states of top pairs and missing transverse energy is presented in Chap. 9. It considers 13.2 fb^{-1} of pp collision data at a centre-of-mass energy of 13 TeV and is interpreted considering two decay scenarios of supersymmetric top quark partners and regarding Dark Matter production in association with top quarks.

References

1. J.W. Goethe, *Faust. Der Tragödie Erster Teil* (Reclam Verlag, Stuttgart, 1986)
2. I. Kant, *Kritik der reinen Vernunft* (Philosophische Bibliothek 505. Felix Meiner Verlag GmbH, Hamburg, 2010)

Chapter 2
The Standard Model

Formulated in the 1960s, the Standard Model of particle physics describes subatomic particle interactions with remarkable success. Its prediction of the top quark and the Higgs boson were the most recent triumphs. Over several orders of magnitude in production cross section, the measurements performed at the Large Hadron Collider (LHC) at CERN precisely confirm the predictions of the Standard Model, as can be seen in Fig. 2.1.

The Standard Model is formulated as a quantum field theory with an underlying $SU(3)_C \times SU(2)_L \times U(1)_Y$ symmetry. $SU(3)_C$ is the gauge group of quantum chromodynamics (QCD) which describes the strong force. Its eight generators $G_\mu^\alpha, \alpha = 1, \ldots, 8$ correspond to the massless spin-1 force carriers, the *gluons*. Particles that are charged under this symmetry are said to carry *colour*, hence the subscript "C". Due to the non-abelian character of $SU(3)_C$, the force carriers themselves are coloured, leading to gluon self-interactions.

$SU(2)_L$ acts exclusively on particles with left-handed chirality, hence the subscript "L". The associated charged-current weak interaction is therefore maximally parity-violating. $U(1)_Y$ is responsible for the *hypercharge* Y. The four generators of $SU(2)_L \times U(1)_Y$ lead to four gauge bosons, denoted as $W_\mu^a, a = 1, 2, 3$ for $SU(2)_L$ and B_μ for $U(1)_Y$. The neutral component of the triplet W_μ^a and the B_μ mix to form the physical degrees of freedom of the neutral bosons: the photon, the force carrier of the electromagnetic force, and the Z, mediating the neutral current of the weak interaction. The two $W^\pm$ bosons, responsible for the charged-current weak interaction, complete the list.

Apart from the spin-1 fields corresponding to the gauge bosons seen above, the Standard Model contains fermions, which are particles with half-integer spin. Coloured fermions are called quarks, while fermions without colour charge are named leptons. They appear in three sets, called *generations*. The gauge interactions of each generation are identical,[1] but the masses of the corresponding fermions

---

[1] Parameters can vary between generations.

© Springer International Publishing AG, part of Springer Nature 2018
J. Gramling, *Search for Dark Matter with the ATLAS Detector*,
Springer Theses, https://doi.org/10.1007/978-3-319-95016-7_2

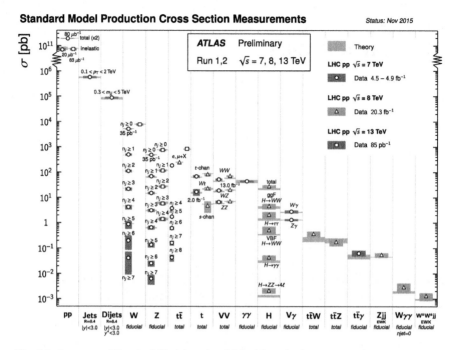

Fig. 2.1 Summary of Standard Model total and fiducial production cross section measurements, compared to the corresponding theoretical expectations from the Standard Model. All theoretical expectations were calculated at next-to-leading-order (NLO) or higher [1]

increase when going from the first to the second to the third generation.[2] Only the light particles of the first generation are stable and make up all matter around us: *up* and *down* quarks form protons and neutrons, the building blocks of atomic nuclei, that are bound with electrons into atoms.

Furthermore, the theory contains an additional complex scalar field, the Higgs field. It was introduced to formulate a mechanism to account for the experimentally observed masses of the electroweak gauge bosons and fermions via spontaneous symmetry breaking. Its corresponding particle, the Higgs boson, was recently discovered by the ATLAS and CMS experiments at the LHC [2, 3].

An overview of the particle content of the Standard Model and its interactions is shown in Fig. 2.2. The Standard Model only contains a description of three of the known four fundamental forces. Gravity, which is much weaker at short scales than the strong, the weak and the electromagnetic forces, is not included.

---

[2]The mass hierarchy of the neutrinos is not yet established and might present an exception.

2.1 Symmetries

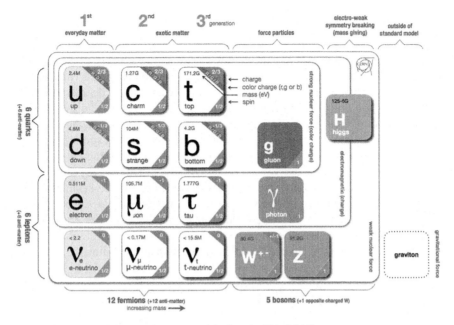

Fig. 2.2 Summary of the particle content of the Standard Model [4]

2.1 Symmetries

Everywhere in physics, symmetry principles are crucial to formulate and better understand the theories of nature. Also the Standard Model is founded on symmetries, characterising and defining its ingredients. According to Noether's theorem, each continuous symmetry is connected with a conserved quantity. Spacetime symmetry implies energy-momentum conservation, conserved charges are the consequence of an exact global symmetry. If a global symmetry is spontaneously broken, massless scalar modes, so-called Goldstone Bosons appear. Exact local symmetries correspond to interactions mediated by a massless spin-1 particle (or spin-2, in the case of gravity). An example is the U(1) symmetry of electromagnetism and its force carrier, the photon. Local broken symmetries relate to interactions by massive spin-1 particles, as it is the case for electroweak interactions. The spontaneous symmetry breaking by the Higgs vacuum expectation value leads to massive force carriers, the W and Z bosons via the so-called Higgs mechanism.

The Standard Model has *imposed* symmetries, namely Lorentz invariance and the above mentioned gauge group $SU(3)_C \times SU(2)_L \times U(1)_Y$. These are, to some extent, motivated by observation. There are also accidental symmetries that either follow from one of the imposed symmetries, the particle content or the requirement of renormalisability. These lead to worthy predictions for experiments and allow one to scrutinise the model in detail.

2.2 Strong Interactions

By imposing a local $SU(3)$ symmetry, one can construct a theory of right-handed and left-handed fermions, the *quarks*, appearing in six colour triplets (fundamental representation, **3** of $SU(3)_C$): $\psi_i(3); i = 1, \ldots, 6$, where *colour* denotes the charge associated to the $SU(3)$ symmetry. This theory is called *Quantum Chromodynamics* (QCD) and describes the strong interactions. Such a theory is straight-forward to define but difficult to solve due to its non-abelian character and its strong coupling in the confinement limit, which constrains the range in which perturbation theory can be applied.

There are eight Lie group generators associated to the SU(3) symmetry, leading to eight gauge bosons–massless spin-1 force carriers, the *gluons* (being in the adjoint representation, **8** of $SU(3)_C$). One single coupling constant, g_s determines the strength of the interaction.

The QCD part of the Standard Model Lagrangian takes the following form:

$$\mathcal{L}_{QCD} = -\frac{1}{4} G^{a\mu\nu} G^a_{\mu\nu} + i \bar{\psi}_i \gamma^\mu D_\mu \psi_i \tag{2.1}$$

with

$$G^a_{\mu\nu} = \partial_\mu G^a_\nu - \partial_\nu G^a_\mu - g_s f_{abc} G^b_\mu G^c_\nu \tag{2.2}$$

and

$$D_\mu = \partial_\mu + i g_s T^a G^a_\mu. \tag{2.3}$$

Here, G^a_μ are the gluon fields, f_{abc} is the structure constant of $SU(3)$, and T_a are the generators of the symmetry, the Gell-Mann matrices. The theory leads to two types of interactions, namely to gluon-fermion vertices and to gluon self-interaction that arises due to the last term in Eq. 2.2:

- Gluon-fermion vertex: $-g_s \bar{\psi} T^a \gamma_\mu G^{a\mu} \psi$
- Gluon self-interaction vertices: $g_s f_{abc}(\partial^\mu G^{a\nu}) G^b_\mu G^c_\nu$ and $g_s^2 f_{abc} f_{ade} G^{b\mu} G^{c\nu} G^d_\mu G^e_\nu$.

These interactions are vectorial, parity-conserving (left/right-handed symmetric), diagonal (no quark mixing terms) and universal (the same coupling appears for all quarks). These predictions, consequences of the imposed symmetry and the consequentially constructed Lagrangian, have been successfully tested by experiments.

The theory also has an accidental, global chiral symmetry, meaning that left- and right-handed particles exhibit independent transformations. This can be used to write:

$$G^{global}_{QCD} = SU(3)_L \times SU(3)_R \times U(1)_A \times U(1)_B. \tag{2.4}$$

The chiral symmetry $SU(3)_L \times SU(3)_R$ is spontaneously broken by the formation of chiral condensates in the QCD vacuum into $SU(3)_F$, the so-called "flavour SU(3)".

The axial symmetry $U(1)_A$, incorporating the part of left-handed particle transformations which are the inverse of the right-handed particle transformations, is classically exact, but broken on the quantum level, which is called an "anomaly". The exact vector symmetry $U(1)_B$, governing the transformations where left- and right-handed particles are treated equally, is identified with baryon number conservation. While $U_B(1)$ is exact independently of the quark masses, $SU(3)_F$ only holds for the light flavour quarks u, d, and s, where they can be assumed to be almost massless. This leads to the approximate isospin symmetry. The accidental global chiral symmetry can also be written as $U(1)_u \times U(1)_d \times U(1)_s \times \cdots$. As a consequence, the quarks are stable and do not decay via the strong interaction.

These accidental symmetries lead to very strong predictions (baryon number conservation, isospin symmetry, stability of quarks w.r.t strong interaction) that have been and continue to be tested in experiments.

The coupling g_s is the only parameter of QCD within the Standard Model. The strong coupling constant $\alpha_s = g_s^2/4\pi$ is determined to be $\alpha_s(m_Z^2) = 0.1182(12)$ [5] at the Z boson mass.[3] The coupling strength depends on the energy scale of the interaction, as described by the renormalisation group equation. It decreases for scales corresponding to high momentum transfers or small length scales, which is known as *asymptotic freedom*. If the distance between two particles is increased, the coupling strength increases as well (almost linearly), until the energy for further separating the particles is sufficient to produce new particles from the vacuum. As a consequence, coloured particles are always bound into colour-neutral objects, hadrons–they are said to be *confined*.

While calculations within the perturbative regime of QCD are able to reach impressive precision, the non-perturbative regime can only be calculated on discrete points of a space-time lattice ('lattice QCD') or phenomenologically described by, for example, hydrodynamic models, in the case of many particles.[4] Heavy-ion collisions studied e.g. at the LHC can help to better understand this regime by characterising hot and dense QCD matter as well as the transition between the phase in which quarks are confined within hadrons and the deconfined Quark-Gluon-Plasma phase.

2.3 Electroweak Interactions

Imposing a local $SU(2)_L \times U(1)_Y$ symmetry leads to a theory of fermions in $SU(2)_L$ doublets of left-handed particles, $Q_{L,i=1,2,3}$ for the quarks and $L_{L,i=1,2,3}$ for the leptons, and in $SU(2)_L$ singlets of right-handed particles, $u_{R,i=1,2,3}$, $d_{R,i=1,2,3}$ and

---

[3]In some cases the so-called *QCD phase* is added to the list of Standard Model parameters of the strong sector as well. It appears in an additional term in the Lagrangian that should be added in general, but since this phase is measured to be very close to zero, this term is neglected here and in many other places. It will be discussed further in Sect. 2.5.

[4]Furthermore, techniques such as QCD sum rules relate hadronic parameters like masses, couplings or magnetic moments, to characteristics of the QCD vacuum, i.e. quark and gluon condensates. Also the so-called quark-hadron duality allows to describe observed reactions either as interactions between partons or of hadronic resonances.

$e_{R,i=1,2,3}$ for up-type quarks, down-type quarks and charged leptons, respectively.[5] The $SU(2)_L$ doublets have a $U(1)_Y$ hypercharge $Y = -1$, the $SU(2)_L$ singlets have a $U(1)_Y$ hypercharge $Y = -2$. A spontaneous breaking of $SU(2)_L \times U(1)_Y \to U(1)_{EM}$ is considered, with $Q_{EM} = I + \frac{1}{2}Y$, where I is the *weak isospin*, the charge of $SU(2)_L$. This requires to extend the theory by a scalar $SU(2)$ doublet, ϕ, with hypercharge $Y = 1$. The four generators, three for $SU(2)$ and one for $U(1)$ result in four gauge bosons, $W^\mu_{a,i=1,2,3}$ and B^μ, all without hypercharges. Two coupling constants determine the strengths of the interactions, g for $SU(2)_L$ and g' for $U(1)_Y$.

The kinetic part of the Lagrangian for fermions and vector bosons takes the following form:

$$\mathcal{L}_{kin} = -\frac{1}{4} W^{i\mu\nu} W^i_{\mu\nu} - \frac{1}{4} B^{\mu\nu} B_{\mu\nu} \qquad (2.5)$$
$$+ i\bar{Q}\gamma^\mu D_\mu Q + i\bar{u}\gamma^\mu D_\mu u + i\bar{d}\gamma^\mu D_\mu d + i\bar{L}\gamma^\mu D_\mu L + i\bar{e}\gamma^\mu D_\mu e,$$

with

$$D_\mu = \partial_\mu - ig W^a_\mu T^a - \frac{i}{2} g' B_\mu. \qquad (2.6)$$

The generators of SU(2) are given by $T^a = \frac{1}{2}\sigma^a$, where σ^a denote the Pauli matrices.

2.3.1 Spontaneous Symmetry Breaking

The model, as described up to now, does not allow the gauge bosons to have mass. However, W and Z bosons, mediating the weak interactions, are measured to be massive. In order to account for that, a scalar $SU(2)$ doublet is introduced. The part of the Lagrangian governing the scalar kinematics and its potential reads:

$$\mathcal{L}_\phi = (D_\mu \phi)^\dagger (D^\mu \phi) - \mu^2 \phi^\dagger \phi - \lambda(\phi^\dagger \phi)^2, \qquad (2.7)$$

where λ is dimensionless, real and needs to be positive to make the potential be bounded from below, and μ^2 has mass dimension two and is assumed to be negative. In this case, the potential has a local maximum at the origin and degenerate minima on a circle around it, satisfying $\phi^\dagger \phi = -\mu^2/(2\lambda)$ and $v^2 = -\mu^2/\lambda$ can be defined. By an $SU(2)$ gauge transformation, a particular vacuum expectation value (vev) of ϕ can be chosen: $\phi_0 = \frac{1}{\sqrt{2}} \binom{0}{v}$. This presents the spontaneous symmetry breaking (SSB): one specific solution that minimises the potential is chosen. When considering small excitations around the vacuum state, ϕ reads:

$$\phi_{SSB}(x) = \frac{1}{\sqrt{2}} \binom{0}{v + h(x)}. \qquad (2.8)$$

---

[5]Note that right-handed neutrinos are not included here.

2.3 Electroweak Interactions

The scalar field $h(x)$ is the only one remaining from the original four fields of ϕ after the SSB. It is identified with the Higgs boson. Substituting ϕ with its vev, the kinetic part of the scalar Lagrangian written in (2.7) can be expressed as:

$$(D_\mu \phi)^\dagger (D^\mu \phi) \supset \left| (g W_\mu^a T^a + g' B_\mu Y) \cdot \begin{pmatrix} 0 \\ v \end{pmatrix} \right|^2 \quad (2.9)$$

$$= \frac{v^2}{4} \left(g^2 (W_\mu^1 W^{1\mu} + W_\mu^2 W^{2\mu}) + (g W_\mu^3 - g' B_\mu)(g W^{3\mu} - g' B^\mu) \right). \quad (2.10)$$

The first part can be written as $m_W^2 W^+ W^-$ with $W_\mu^\pm = \frac{1}{\sqrt{2}}(W_\mu^1 \mp i W_\mu^2)$ and hence leads to the mass terms for the charged W bosons:

$$m_W^2 = v^2 g^2 / 4. \quad (2.11)$$

The second term of Eq. 2.10 leads to the mass of the neutral Z boson with $Z_\mu = W_\mu^3 \cos\theta_W - B_\mu \sin\theta_W$, where θ_W is the so-called Weinberg angle with $\tan\theta = g'/g$:

$$m_Z^2 = v^2 (g^2 + g'^2)/4. \quad (2.12)$$

The neutral counterpart of the Z boson, the photon (A), remains massless:

$$A_\mu = W_\mu^3 \sin\theta + B_\mu \cos\theta_W. \quad (2.13)$$

The mass eigenstates Z and A are formed by W^3 and B as:

$$\begin{pmatrix} Z \\ \gamma \end{pmatrix} = \begin{pmatrix} \cos\theta_W & -\sin\theta_W \\ \sin\theta_W & \cos\theta_W \end{pmatrix} \begin{pmatrix} W^3 \\ B \end{pmatrix} \quad (2.14)$$

A real, massive scalar degree of freedom, the Higgs boson, whose mass is given by $m_h^2 = 2\lambda v$ results from the SSB. It has been measured to be $m_H = 125.09 \pm 0.24$ GeV [5].

2.3.2 Fermion Masses

Without SSB, it is not possible to write down mass terms for the fermions. They are *chiral*: left-handed and right-handed particles behave differently, since only left-handed particles are charged under $SU(2)_L$. This excludes a Dirac mass term in which left- and right-handed particles ought to be combined and so would break $SU(2)_L$. Furthermore, a Majorana mass term is excluded, since the fermions are charged and hence cannot be their own antiparticles.

However, a Yukawa interaction between fermions and the complex scalar field ϕ can be written down in the following way:

$$\mathcal{L}_{Yuk} = Y_{ij}^d \bar{Q}_{L_i} \phi d_{R_j} + Y_{ij}^u \bar{Q}_{L_i} \tilde{\phi} u_{R_j} + Y_{ij}^l \bar{L}_{L_i} \phi e_{R_j} + h.c.. \quad (2.15)$$

The matrices Y_{ij} contain the different Yukawa coupling strengths. Note that the up-type quarks couple to $\tilde{\phi} = -i\sigma_2 \phi*$. After spontaneous symmetry breaking, i.e. the transition $\phi \to \binom{0}{v}$ (and $\tilde{\phi} \to \binom{v}{0}$) the above Yukawa terms have the form of a Dirac mass term with the fermion mass given by $m_f = v y_{ij}/\sqrt{2}$, where v denotes the Higgs vev and y_{ij} the relevant Yukawa coupling. Since the Standard Model does not contain right-handed neutrinos, mass terms for neutrinos are not possible and hence they remain massless and degenerate.

For leptons, the interaction basis can always be made consistent with the mass basis, such that Y_{ij}^l is diagonal. This is not the case for the quarks: generally, no interaction basis can be found that is also a mass basis for both up- and down-type quarks, i.e. that diagonalises both Y_{ij}^d and Y_{ij}^u at the same time. Hence, the mass eigenstates of the quarks generally do not coincide with the flavour eigenstates which take part in the electroweak interaction. This leads to a mixing of flavour states to form the mass eigenstates that is described by a unitary matrix, called *CKM* matrix after Cabbibo, Kobayashi and Maskawa. Its complex phase gives rise to CP-violating processes within the Standard Model.

Electromagnetic Interactions

Electromagnetic interactions are vectorial, parity conserving, diagonal and universal:

$$\mathcal{L}_{EM} = eq_f \bar{\psi}_{f_i} \gamma_\mu A^\mu \psi_{f_i}, \quad (2.16)$$

where $e = g \sin \theta_W$ is the electromagnetic coupling, q_f is the electromagnetic charge of the (left- or right-handed) fermion ψ_f and A^μ denotes the photon.

Neutral Currents

Neutral weak interactions, mediated by the Z boson, are chiral, i.e. they distinguish between left- and right-handed particles and hence violate parity. They are diagonal and universal since fermions are in the same representation as for $U(1)_{EM}$.

$$\mathcal{L}_{NC} = \frac{g}{\cos(\theta_W)} \left(I^3 \frac{1-\gamma^5}{2} - q_f \sin^2(\theta_W) \right) \bar{\psi}_{f_i} \gamma_\mu Z^\mu \psi_{f_i}. \quad (2.17)$$

As above, q_f denotes the electromagnetic charge and ψ_{f_i} stands for any fermion. I^3 is the third component of the weak isospin. The projection operator $(1-\gamma^5)/2$ selects only the left-handed components of ψ_{f_i}. This means, the neutral current consists of a purely left-handed part, proportional to I^3 and a part treating left- and right-handed particles equally, proportional to q_f.

2.3 Electroweak Interactions

Charged Currents

Charged weak interactions are mediated by the W bosons. Since they arise purely from $SU(2)_L$, charged currents involve only left-handed fermions and are hence maximally parity-violating. As long as neutrinos are treated massless and degenerate, the lepton interactions are diagonal and universal, while the above-mentioned CKM matrix describes the mixing in the case of quarks, where the charged currents are neither universal nor diagonal.

$$\mathcal{L}_{CC} = -\frac{g}{\sqrt{2}} \left[\bar{\psi}_{L,u_i} \gamma^\mu M_{CKM} \psi_{L,d_i} + \bar{\psi}_{L,\nu_i} \gamma^\mu \psi_{L,l_i} \right] W^+ + h.c.. \quad (2.18)$$

Here, ψ_{L,d_i} and ψ_{L,u_i} stand for down- and up-type quarks, respectively, while ψ_{L,d_i} and ψ_{ν,d_i} indicate charged leptons and neutrinos.

Vector Boson Interactions

In addition, the Lagrangian contains the following terms, covering three- and four-point interactions between the vector gauge bosons:

$$\begin{aligned}\mathcal{L}_{VVV} = -ig[(W^+_{\mu\nu} W^{-\mu} - W^{+\mu} W^-_{\mu\nu})(A^\nu \sin\theta_W - Z^\nu \cos\theta_W) \\ + W^-_\nu W^+_\mu (A^{\mu\nu} \sin\theta_W - Z^{\mu\nu} \cos\theta_W)]\end{aligned} \quad (2.19)$$

$$\mathcal{L}_{VVVV} = -\frac{g^2}{4} \left[(2W^+_\mu W^{-\mu} + (A_\mu \sin\theta_W - Z_\mu \cos\theta_W)^2)^2 \right.$$
$$\left. -(W^+_\mu W^-_\nu + W^+_\nu W^-_\mu + (A_\mu \sin\theta_W - Z_\mu \cos\theta_W)(A_\nu \sin\theta_W - Z_\nu \cos\theta_W))^2 \right]$$
$$(2.20)$$

Higgs Interactions

The Yukawa couplings between the Higgs and the fermions, as introduced above, are proportional to the particle masses, heavier particles couple stronger to the Higgs. The Yukawa couplings are diagonal.

Recall that after SSB the Higgs field ϕ takes the following form: $\phi_{SSB}(x) = \frac{1}{\sqrt{2}} \binom{0}{v+h(x)}$. Dimensionless couplings such as $hhhh$ and $hhVV$ involving only $h(x)$ but not v, arise in the Lagrangian. Trilinear couplings like hhh and hVV vertices are proportional to v. The cubic and quartic Higgs self-interactions are given by:

$$\mathcal{L}_H \supset +\frac{\lambda v}{2} H^3 + \frac{\lambda}{4} H^4. \quad (2.21)$$

The interactions between the Higgs and the vector bosons reads:

$$\mathcal{L}_{HV} \supset \left(g^2 \frac{v}{2} H + \frac{g^2}{4} H^2\right) \left(W_\mu^+ W^{-\mu} + \frac{1}{2\cos^2\theta_W} Z_\mu Z^\mu\right). \tag{2.22}$$

Couplings between the Higgs and the photon are not allowed since the Higgs is not charged electromagnetically and the photon is massless and does not have a Yukawa interaction with the Higgs.

2.4 Defining the Standard Model

Combining what was seen in the sections above, the Standard Model can be defined as a theory with a $SU(3)_c \times SU(2)_L \times U(1)_Y$ symmetry. It contains three colour triplets, the quarks, out of which the left-handed ones can be grouped in $SU(2)_L$ doublets: $Q_{L_i}(3, 2)_{+\frac{1}{6}}$, whereas the right-handed colour triplets are $SU(2)_L$ singlets: $u_{R_i}(3, 1)_{+\frac{2}{3}}, d_{R_i}(3, 1)_{-\frac{1}{3}}$. It also contains colour-singlets, the leptons, coming in left-handed $SU(2)_L$ doublets, $L_{L_i}(1, 2)_{-\frac{1}{2}}$ and right-handed $SU(2)_L$ singlets, $e_{R_i}(1, 1)_{-1}$. Right-handed neutrinos are not considered. The generators of $SU(3)_C$ and $SU(2)_L \times U(1)_Y$ are taken to commute, leading to a theory with three couplings. The 12 generators lead to 12 gauge bosons: eight gluons, the photon, the Z boson and two W bosons. Spontaneous symmetry breaking of $SU(2)_L \times U(1)_Y \to U(1)_{EM}$ is included via a scalar $SU(2)_L$ doublet, whose neutral component, the degree of freedom that persists after SSB, presents the Higgs boson. The SSB results in only $SU(3)_C \times U(1)_{EM}$ remaining unbroken.

In summary the Lagrangian of the Standard Model reads:

$$\mathcal{L} = -\frac{1}{4} G_{\mu\nu} G^{\mu\nu} - \frac{1}{4} W_{\mu\nu} W^{\mu\nu} - \frac{1}{4} B_{\mu\nu} B^{\mu\nu} \tag{2.23}$$

$$+ \bar{\psi}_{L_i} \gamma^\mu \left(i\partial_\mu - g_s T G_\mu - g \frac{\sigma}{2} M_{CKM} W_\mu - g' \frac{Y}{2} B_\mu \right) \psi_{L_i} \tag{2.24}$$

$$+ \bar{\psi}_{R_i} \gamma^\mu \left(i\partial_\mu - g_s T G_\mu - g' \frac{Y}{2} B_\mu \right) \psi_{R_i} \tag{2.25}$$

$$+ \left|\left(i\partial_m u - g \frac{\sigma}{2} W_\mu - g' \frac{Y}{2} B_\mu \right) \phi \right|^2 - V(\phi) \tag{2.26}$$

$$- (Y_{ij}^d \bar{Q}_i \phi d_j - Y_{ij}^u \bar{Q}_i \tilde{\phi} u_i - Y_{ij}^\ell \bar{L}_i \phi e_i + h.c.. \tag{2.27}$$

The different parts govern the gauge boson kinetic terms and self-interactions (Eq. 2.24), the kinetic terms and interactions of the left-handed (Eq. 2.24) and right-handed (Eq. 2.25) fermions, the boson mass terms and couplings (Eq. 2.26) and the coupling terms between the Higgs and the fermions, leading to the fermion masses (Eq. 2.27).

The kinetic part of the Standard Model Lagrangian exhibits a $U(3)_Q \times U(3)_U \times U(3)_D \times U(3)_L \times U(3)_E$ symmetry that gets broken by the Yukawa coupling terms into $U(1)_B \times U(1)_e \times U(1)_\mu \times U(1)_\tau$, representing the symmetries leading to baryon number conservation[6] and lepton family number conservation.

2.4.1 Parameters

Collecting all parameters results in three coupling constants, g_s, g, g', responsible for the strength of the strong, the weak $SU(2)_L$ and the hypercharge $U(1)_Y$ interactions. In addition, there is the mass of the Higgs, m_H and its vacuum expectation value v. Further, three lepton and six quark masses, as well as three CKM angles and one complex CKM phase have no theoretical prediction and need to be measured. This results in 18 parameters in total.[7]

2.5 Open Questions and Known Problems of the Standard Model

2.5.1 Neutrino Masses

The Standard Model only accounts for massless neutrinos. The unambiguous measurements of neutrino oscillations requires the neutrinos to have a mass different from zero [5]. Hence, neutrino masses present physics beyond the Standard Model. To accommodate neutrino masses, the Standard Model can be extended to the so-called νSM. Either, neutrinos are continued to be assumed to be Dirac particles and right-handed neutrinos are proposed. They would be $SU(2)$ singlets and hence would exhibit no Standard Model interactions apart from those involving the Higgs field. Neutrinos would then get Dirac masses as all other fermions:

$$\mathcal{L}_\nu \supset y_{ij}^D L_i \tilde{\phi} \nu_{R_j}. \tag{2.28}$$

Here, y_{ij}^D gives the relevant Yukawa coupling, L denotes the $SU(2)$ doublet of left-handed leptons, ν_{R_j} the $SU(2)$ singlet of right-handed neutrinos and $\tilde{\phi} = -i\sigma_2 \phi *$ where ϕ is the Higgs field. A second way would be to postulate Majorana mass terms for neutrinos, where neutrinos would be their own antiparticles. This would add additional dimension-five operators to the Standard Model of the form:

$$\mathcal{L}_\nu \supset y_{ij}^M / v \phi \phi \nu_i \nu_j. \tag{2.29}$$

---

[6]The Baryon symmetry is broken non-perturbatively by sphaleron transitions [6].
[7]The QCD phase θ, mentioned above and discussed in detail below, sometimes gets counted here as well, resulting in a total number of 19 parameters.

Here, $\tilde{y}_{ij}^M$ gives the couplings that lead the neutrino masses. Such terms violate the lepton and flavour conservation. Furthermore, they are of dimension five and non-renormalisable. Hence, they are only meaningful, if new physics is introduced at a scale Λ_{NP}:

$$\mathcal{L}_{SM} \supset \frac{\tilde{y}_{ij}^M}{\Lambda_{NP}} \phi\phi \nu_i \nu_j \rightarrow \frac{\tilde{y}_{ij}^M v^2}{2\Lambda_{NP}} \nu_i \nu_j. \tag{2.30}$$

With $m_\nu \approx 0.1\,\text{eV}$ and couplings of order one this scale would be as high as $\Lambda_{NP} \sim 10^{14}$.

2.5.2 Fine Tuning and Naturalness

The concept of *naturalness* demands the parameters of a theory to take relative values "of order one", meaning there occur no exceptionally large or small parameters without explanation. Such exceptional parameters could either be realised via choosing very particular small (or large) coefficients to the relevant terms, or some very precisely adjusted, or so-called *fine-tuned*, mechanism to cancel (or enhance) the relevant effects. Within the Standard Model, there occur several naturalness problems as will be discussed in the following.

The notion of naturalness significantly influenced the arguments in modern particle physics during the last decades. The refusal to accept unnatural or fine-tuned parameters within the Standard Model triggered the development of a plethora of models introducing new physics that provide for some mechanism to restore naturalness.

2.5.2.1 Dimension 0: Cosmological-Constant Problem

If an operator of the lowest possible dimension is constructed, no dependence on fields or derivatives is present. It is not forbidden for such dimension-0 operators to appear in the Standard Model Lagrangian, effectively adding an arbitrary constant to the Lagrange density. Commonly, such an operator is identified with the energy density of the vacuum used to renormalise the zero-point energy of the quantum field theory (QFT).

Generally, such an absolute QFT vacuum energy is irrelevant, as typically only energy differences are directly observable in high-energy physics. But it is measurable in a cosmological context, as gravity couples to all forms of energy. The QFT vacuum energy can be identified with Einstein's cosmological constant term in general relativity. A change in this parameter alters the expansion history of the universe. Currently, the energy density from the cosmological constant is measured to be below $\rho_{CC} \sim (3 \times 10^{-3}\,\text{eV})^4$ [5].

Connecting the particle-physics and gravitational ends, it seems extremely puzzling that this value is so small. Contributions to the vacuum energy would be

2.5 Open Questions and Known Problems of the Standard Model

expected to be of the order of $\Delta\rho_0 \sim (\frac{\Lambda^4}{16\pi^2})$. This quantity is much larger than the measured value of the cosmological constant considering the mass of any known particle (except for neutrinos) as Λ. Considering the highest assumed physics scale, the Planck scale M_{Pl}, for Λ results in a difference of 122 orders of magnitude between the naively estimated and the measured value of ρ_{CC}!

2.5.2.2 Dimension 2: Hierarchy Problem

It is important to note the enormous difference between the electromagnetic and the gravitational scale: gravity is characterised by a single, dimensionful quantity, Newton's constant G_N. It defines a very large mass scale $M_{Pl} = G_N^{-\frac{1}{2}} \simeq 1.22 \times 10^{19}$ GeV [5], the Planck scale, which is tens of orders of magnitude higher than the electroweak scale around the Higgs vev. The *hierarchy* between these scales results in another naturalness problem of the Standard Model.

The lowest-dimensional operator involving Standard Model fields is of dimension two and would take the following form:

$$\mathcal{L}_2 = \mu^2 \phi^\dagger \phi \tag{2.31}$$

with $m_H^2 = 2\mu^2$. On the other hand, loop corrections to μ^2 are given by:

$$\Delta(\mu^2) \propto \frac{\Lambda^2}{16\pi^2}. \tag{2.32}$$

Assuming M_{Pl} for Λ (or even the mass scale Λ_{NP} derived from the dimension-5 description of neutrino masses as explained above: 10^{14} GeV) leads to values much larger than the value of μ^2 that is needed to account for the observed value of m_H.[8] Some cancellation between the not directly accessible correction terms might be possible, but it would need to be very accurately tuned, presenting a major fine-tuning problem.

There are several proposals on how to evade this problem:

- Assume, that the EW scale is the highest scale that exists. Obvious problems of this approach, e.g. with M_{Pl} being well motivated, can be avoided by the introduction of extra dimensions.
- Assume a composite model in which the Higgs is not elementary. Hence, no elementary scalar particle would exist and the problematic corrections do not apply anymore. Typically, such models involve additional gauge interactions, similar to QCD (e.g. *technicolor*).
- Assume additional symmetries that either provide some cancellation of the correction terms to μ^2 (e.g. Supersymmetry relating bosons and fermions) or make the

---

[8]Note that this is technically only true in the presence of new, heavy states at the scale Λ.

Higgs a pseudo-goldstone boson (neutral naturalness) or protect the small Higgs mass via technical naturalness.
- More recent proposals include also the introduction of a so-called *Relaxion*, which relates the Higgs-mass problem to inflation, or *N-Naturalness*, where several copies of the Standard Model are assumed.

2.5.2.3 Dimension 4: Strong CP Problem

Without being necessary for gauge invariance, but also not forbidden by any of the imposed symmetries, an additional term can be–and generally should be–added to the QCD Lagrangian:
$$\mathcal{L}_{QCD} \supset \theta_{QCD} \epsilon_{\mu\nu\rho\sigma} G^{\mu\nu} G^{\rho\sigma}. \tag{2.33}$$

This introduces CP violation in the strong sector. However, measurements such as the one of the neutron electric-dipole moments suggest that $\theta_{QCD} \lesssim 10^{-10}$. It is not understood why this parameter is so small, or respectively why this term should not appear in the Standard Model Lagrangian. One approach is to postulate an additional symmetry (the so-called Peccei-Quinn symmetry) that forbids such a term, which comes with the presence of new particles, the QCD axions.

2.5.3 Further Questions

2.5.3.1 Triviality Problem

The Standard Model couplings change with the scale at which they are evaluated. Whereas the strong coupling vanishes at high energies (known as *asymptotic freedom*) the hypercharge gauge coupling (g') increases with increasing energy scale and gets infinitely large at finite energies (at the *Landau Pole*), which of course marks a breakdown of the theory. Such theories are named *trivial* since they only make sense if the coupling in question is zero and hence the theory is trivial. This feature is however no problem in case one assumes that the theory is only a low-energy description and still lacks a UV completion, but is of course another manifestation of the view that the Standard Model needs to be amended to form a larger theory.

2.5.3.2 Flavour Problem

The Standard Model offers a description of the mechanism of how fermions (apart from neutrinos) obtain mass and how they mix, but needs a large number of input parameters to do so. It does not explain the peculiar observed structure or hierarchy in mass of the different generations.

2.5.3.3 Matter-Antimatter Asymmetry

Within the Standard Model, matter and antimatter are produced in almost equal amounts. The CP violation that is introduced by the imaginary phase of the CKM matrix is by far not sufficient to explain the fact that the universe seems to be almost exclusively made up by matter. To explain this, additional mechanisms of baryo- and leptogenesis have to be assumed.

2.5.3.4 Dark Matter

There are many cosmological and astrophysical observations providing evidences for a new type of very weakly-interacting matter, Dark Matter. However, the Standard Model does not contain a plausible candidate for Dark Matter. This will be discussed in the next chapter.

References

1. ATLAS Collaboration, Standard model production cross section measurements (2013), https://atlas.web.cern.ch/Atlas/GROUPS/PHYSICS/CombinedSummaryPlots/SM/ATLAS_b_SMSummary_FiducialXsect/ATLAS_b_SMSummary_FiducialXsect.png
2. ATLAS Collaboration, G. Aad et al., Observation of a new particle in the search for the Standard Model Higgs boson with the ATLAS detector at the LHC. Phys. Lett. **B716**, 1–29 (2012). https://doi.org/10.1016/j.physletb.2012.08.020, arXiv:1207.7214 [hep-ex]
3. CMS Collaboration, S. Chatrchyan et al., Observation of a new boson at a mass of 125 GeV with the CMS experiment at the LHC. Phys. Lett. **B716**, 30–61 (2012). https://doi.org/10.1016/j.physletb.2012.08.021, arXiv:1207.7235 [hep-ex]
4. A. Purcell, Go on a particle quest at the first CERN webfest. Le premier webfest du CERN se lance la conque des particules, https://cds.cern.ch/record/1473657
5. C. Patrignani, Review of particle physics. Chin. Phys. **C40**(10), 100001 (2016). https://doi.org/10.1088/1674-1137/40/10/100001
6. G.R. Farrar, M.E. Shaposhnikov, Baryon asymmetry of the universe in the minimal standard model. Phys. Rev. Lett. **70**, 2833–2836 (1993). https://doi.org/10.1103/PhysRevLett.70.2833

Chapter 3
Dark Matter

Around 1930, the first observations inconsistent with the assumption that ordinary, visible matter is all there is in the universe were made. After evaluating less radical solutions, scientists proposed a new type of matter that would only interact gravitationally and at most weakly with ordinary matter: *Dark Matter*. However, the study of a part of the universe that can hide so well, that is coexisting with our world without interacting with it, is difficult. Nevertheless, it is extremely relevant: there exists about five times more Dark Matter (DM) than ordinary matter in the universe–even if the Standard Model and its open questions would be fully understood, it would only concern a small fraction of the content of the world. Furthermore, DM played a crucial role in the history of the universe, also for our galaxy and solar system. In the following, a selection of empirical evidence for DM is given in Sect. 3.1 before the inferred properties of DM are discussed in Sect. 3.2. The often assumed scenario of DM as a thermal relic is outlined in Sect. 3.3. Possible candidates for DM are discussed in Sect. 3.4 before Sect. 3.5 gives an overview over DM searches. Finally, Sect. 3.6 gives an introduction to the effective field theoretic models used to describe DM production at the LHC.

3.1 Evidence for Dark Matter

The fact that evidence for DM was found on very different scales disfavoured alternative proposals and eventually led to the assumption of this new component of matter. The most important experimental observations substantiating the proposal of DM are presented in the following.

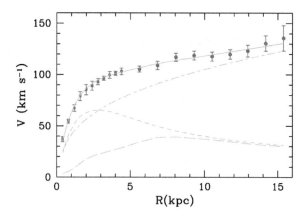

Fig. 3.1 M33 rotation curve (points) compared with the best fit model (continuous line). The different components of the best fit model are shown as well: the DM halo (dashed-dotted line), stellar disk (short dashed line), and gas contributions (long dashed line). Figure from Ref. [4]

3.1.1 Galactic Scales: Galaxy Rotation Curves

Fritz Zwicky coined the term *Dark Matter*. In his studies of the Coma galaxy cluster and the Virgin Cluster, he observed rotation velocities of galaxies of up to 400 times higher than what he expected. This astounding observation led him to propose a new form of additional gravitational matter present in galaxy clusters, Dark Matter [1]. Soon after, in 1939, the American astronomer Horace W. Babcock studied the rotation velocities of light-emitting objects of the Andromeda nebula and observed that the mass-to-light ratio differed significantly from his expectations, resulting in an almost constant velocity over distance from the galaxy centre [2] (such as in Fig. 3.1). He attributed this observation to the underestimation of the luminous matter due to additional light absorption mechanisms taking place inside the galaxy. Zwicky's proposal of Dark Matter was revived only in 1970, when his measurements were repeated with similar outcomes [3].

The rotational velocity of an orbiting object at distance r scales like $v(r) \propto \sqrt{M(r)/r}$ where $M(r)$ is the mass enclosed by the orbit. Outside the (visible) galaxy one would hence expect: $v(r) \propto 1/\sqrt{r}$. But v was found to be approximately constant out to large values of r as can be exemplarily seen in Fig. fig:galaxyrotation. This implies $M(r) \propto r$, which is not observed in luminous matter. Hence, the existence of a dark halo with mass density $\rho(r) \propto 1/r^2$ is proposed.[1]

The question of how exactly DM is distributed within the halo is not yet fully settled. There are several proposed density profiles that are well-motivated and match the observations. They are discussed in detail in Appendix A.1. The choice of a DM halo density profile presents one substantial uncertainty that enters the expectation of DM annihilation or DM-nucleon scattering.

---

[1] At some point ρ would need to drop off faster in order to arrive with a finite mass of the galaxy.

3.1.2 Cluster Scales: Gravitational Lensing

Following Einstein's laws of gravity, light rays are deflected by gravitational objects, since they travel on straight lines in space and massive objects curve the space itself. This effect can be used to infer masses of objects between a light-emitting object and an observer. The deflection angle, θ, depends on the mass of the object, M, and its distance to the source of light, r, in the following way:

$$\theta = \frac{4G_N M}{rc^2}, \tag{3.1}$$

where G_N is the gravitational constant and c the speed of light.

Depending on the size of the effect, three kinds of gravitational lensing are distinguished. *Strong gravitational lensing* leads to easily visible distortions such as multiple images, arcs and full rings, so-called Einstein rings. *Weak gravitational lensing* is only detectable via statistical analysis of many sources, which tend to be stretched perpendicular to the lensing object. The shape, size and orientation of the light-emitting sources can be used to determine the mass of the lensing object. This means in particular that the component of DM present in the lensing object (e.g. a galaxy cluster) can be determined. In *microlensing* the effect is even smaller and can only be detected as an apparent change of brightness of the source.

By studying weak lensing effects, the merger of two galaxy clusters, one of which is named the *Bullet Cluster*, was observed at a distance of 3.8 billion lightyears (see Fig. 3.2). This observation is particularly relevant in view of DM. First, an unambiguous spatial offset between the total centre of mass and the centre of baryonic mass strongly supports the hypothesis of an additional, non-baryonic matter component present in galaxy clusters and hence is direct evidence for DM on large cluster scales. Secondly, it disfavours theories of modified gravity (MONDs), and favours particle DM models. Furthermore, it constrains the self-interaction of DM.

3.1.3 Cosmological Scales: Cosmic Microwave Background

A crucial point in the history of the universe is the start of the so-called recombination phase. At this point, the temperature of the universe was small enough for charged particles to move sufficiently slow to allow them to be bound together into neutral objects–atoms. As a consequence, photons no longer scattered on charged particles and hence started to travel basically unhindered.

The possibility to observe these photons as "background radiation" was already predicted in 1948 by Alpher and Herman [6]. The first observation of these remnant photons was made accidentally by Penzias and Wilson in 1965, when they noticed a uniform background radiation in their measurement devices designed for radio astronomy [7]. The uniform radiation corresponded to a temperature of 4.2 K (the value was later corrected to 2.7 K). From the binding energy of the Hydrogen atom,

Fig. 3.2 The galaxy cluster 1E 0657-56, also known as the *Bullet Cluster*, which was formed after the collision of two large clusters of galaxies as seen by the *Chandra* telescope. Hot gas, detected in X-rays, is coloured in pink, containing most of the baryonic matter. An optical image from the *Magellan* and the *Hubble Space Telescope* shows the galaxies in orange and white. The blue colour indicates the mass distribution within the clusters, determined via weak gravitational lensing. Most of the matter in the clusters (blue) is clearly separated from the luminous matter (pink), giving direct evidence that a significant amount of the matter in the clusters is dark. Figure from Ref. [5]

one would expect a temperature around 3740 K,[2] however, the expansion of the universe shifted the wavelength of the photons by $(1 + z)$, where z is the red-shift related to the expansion rate of the universe.

The first striking observation is that this *Cosmic Microwave Background* (CMB) is very uniform: dipole fluctuations are of the order of 10^{-3} K, further fluctuations are below 10^{-5} K. These temperature fluctuations are explained as the result of small, primordial density fluctuations, leading to acoustic oscillations in the photon-baryon plasma present before recombination. The angular scales of CMB oscillations, measured as the power spectrum of the anisotropies, reveal the different effects of baryonic and DM. While ordinary matter interacts strongly with radiation, DM does not. Hence, these components affect the plasma oscillations differently.

The anisotropies can be parametrised as follows:

$$\langle \Delta T^2 \rangle = \left(\frac{l(l+1)}{2\pi} C_l \right) \langle T \rangle^2, \qquad (3.2)$$

where l are the multipole moments and C_l the angular frequencies. The acoustic spectrum of the CMB, measured by the *Planck* satellite, is shown in Fig. 3.3. The spectrum shows a large first peak and smaller successive peaks. It is sensitive to the

---

[2]This value is calculated from the binding energy of the hydrogen atom, 13.6 eV or 157760 K, considering the Boltzmann distribution of energies of the much more abundant photons.

3.1 Evidence for Dark Matter

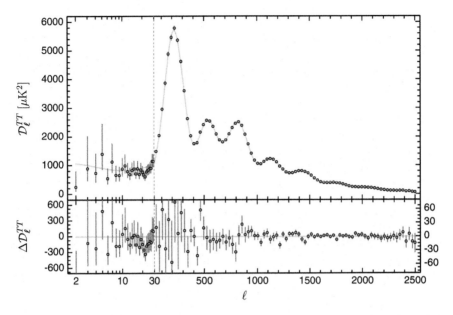

Fig. 3.3 The *Planck 2015* temperature power spectrum. The best-fit based ΛCDM theoretical spectrum fitted to the *Planck* data is plotted in the upper panel. Residuals with respect to this model are shown in the lower panel. The error bars show ±1σ uncertainties. Figure from [8]

DM density in the following ways: enhancing the DM density reduces the overall amplitude of the peaks and lowering the DM density significantly reduces the amplitude of the third peak, due to the smaller coupling between DM to radiation. The observed spectrum presents a clear third maximum and is consistent with the DM hypothesis. Hence, evidence for DM is also found on cosmological scales.

The careful analysis of the acoustic spectrum of the CMB constrains the relative abundance of the different components of the universe. The *Planck* satellite provides the most accurate measurement to date and indicates that only 4.9% of the energy content of the universe is accounted for by ordinary matter, 26.8% by DM and 68.3% by so-called Dark Energy, possibly identified with a cosmological constant [8]. Based on these and other astrophysical and cosmological measurements, the so called *Standard Model of Cosmology* (ΛCDM) was formulated, in which the presence of DM, as well as Dark Energy plays a key role. It is described in detail in Appendix A.3.

3.2 Properties of Dark Matter

Apart from the above mentioned measurements, large-scale structures, big-bang nucleosynthesis, and other observations have constrained the properties of DM. Commonly, it is assumed that DM is made up of particles. Under this assumption, basically all observations, on all scales, can be reproduced. However, there are also several models proposed that try to explain the evidence for DM in different ways. One alternative is to assume that DM is built up of ordinary matter, but it is clumped together in MAssive, Compact Halo Objects (MACHOs), preventing it from interacting with other baryonic matter. Candidates for such MACHOs include (primordial) black holes, neutron stars, and brown dwarfs. Such a proposal is severely challenged by the hints of the non-baryonic nature of DM coming from the CMB analysis and big-bang nucleosynthesis. In addition, searches for MACHOs span by now almost the entire mass range allowed for an explanation of DM phenomena, leaving only a small window open [9]. Another alternative to particle DM would be to assume a modification to general relativity [10]. Although there exist relativistic theories that successfully reproduce some of the observed phenomena, like galaxy rotation curves, other measurements, for example those relating to properties of galaxy clusters, cannot be easily explained by such models [11]. It is in general not easy to formulate a theory of modified gravity that can be incorporated in any cosmological model [12].

For the rest of this thesis, a scenario of particle DM is adapted. Summarising the different observations, the following properties of particle DM are most generally assumed, although, for many aspects, there are also models proposed that circumvent these assumptions:

- It is *dark*: electrically neutral and colour neutral
- It is *cold*: large-scale structure formation would be altered and inconsistent with observations, if DM was relativistic at the time of matter-radiation equality
- It is non-baryonic
- It interacts gravitationally
- Its self-interaction cannot be too strong
- If it interacts with ordinary matter, then it only does so weakly
- It makes up 26.8% of the universe's mass-energy content, around five times more than ordinary matter
- It is sufficiently long-lived, given that there is a non-vanishing relic density today.

The only potential candidate within the Standard Model is the neutrino. However, neutrinos would correspond to *hot* DM, since they moved with relativistic velocities during the epoch of matter-radiation equality and would lead to an altered structure formation. In addition, they cannot account for the observed DM density as will be seen in the following. Hence, DM has to come from new physics, most probably from one (or many) new particle(s). But the scale of masses and interactions of DM is largely unknown.

3.3 Thermal Dark Matter

Dark Matter is often assumed to be a thermal relic, meaning that, in the early universe, its particles were in thermal equilibrium. At some point the expansion rate of the universe exceeded the total interaction rate of DM, and since then it exhibited

3.3 Thermal Dark Matter

basically no interactions. At this point, it dropped out of thermal equilibrium and *decoupled*. The decoupling time and temperature, also called *freeze-out* temperature, determine the relic abundance of DM today in the following way.

First, the Boltzmann equation relates the particle number density n to the interaction strength:

$$\frac{dn}{dt} + 3H_0 n = -\langle \sigma v \rangle (n^2 - n_{eq}^2), \qquad (3.3)$$

where $\langle \sigma v \rangle$ is the thermal average of the total DM annihilation cross section multiplied by the velocity, H_0 is the Hubble constant ($H_0 = 678.9 \text{ km s}^{-1}$ [13]) and n_{eq} is the number density in thermal equilibrium. The latter can be determined in the non-relativistic limit as:

$$n_{eq} = g \left(\frac{mT}{2\pi} \right)^{3/2} e^{-m/T}, \qquad (3.4)$$

with g giving the number of degrees of freedom, m being the particle mass and T the temperature.

The Boltzmann equation (Eq. 3.3) can then be solved for times long before and long after the freeze-out. Matching the solution yields a value for the relic density. Based on this strategy, the following often-used order-of-magnitude estimation can be motivated:

$$\Omega_\chi h^2 \approx \frac{3 \times 10^{-27} \text{cm}^3 \text{s}^{-1}}{\langle \sigma v \rangle}. \qquad (3.5)$$

The relic density is expressed relative to the critical density as $\Omega_\chi = \rho_\chi / \rho_c$ with $\rho_c = 3H_0^2 / 8\pi G_N$, and h denotes the Hubble parameter $h = H_0 / 100 \text{ km s}^{-1} \text{ Mpc}^{-1}$. The evolution of the co-moving number density with time is sketched in Fig. 3.4. Smaller interaction cross sections correspond to earlier freeze-out temperatures and therefore higher relic abundances and vice versa. It has to be noted that this simple estimate is only approximate and can be severely altered, for example when coannihilations with particles that are only slightly heavier than the stable DM particle can occur.

The experimentally determined value of the relic density of DM is extracted from a global fit to several measurements, for example the CMB results of the *Planck* satellite [15]:

$$\Omega_\chi h^2 = 0.1186(20). \qquad (3.6)$$

Interestingly, the measured relic abundance can be obtained when assuming a DM mass and interaction similar to the weak scale. Such a weakly interacting massive particle is called a WIMP. This presents an unlikely coincidence: two scales, the relic abundance of DM and weak scale parameters, seem to be connected. This goes under the name *WIMP miracle* and motivates looking for DM candidates with masses between roughly 1 GeV and 1 TeV and interactions with the Standard Model sector of the order of the weak interaction. Such a scenario could be detected at the LHC or by other DM search programs.

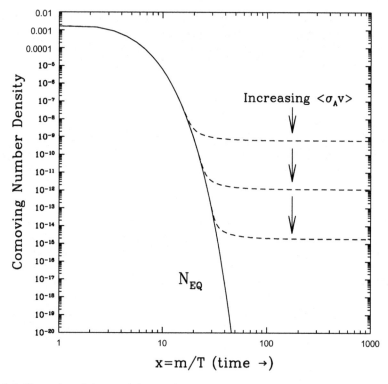

Fig. 3.4 Illustration of the comoving number density of a stable particle species as it evolves through the process of thermal freeze-out [14]

3.4 Candidates for Non-baryonic Dark Matter

The presented gravitational observations of DM as well as the estimated relic density are consistent with a variety of DM masses and interaction strength. Hence, it should not be forgotten that the landscape of possible candidates is large and any search for DM necessarily only probes a small subset of proposed candidates. In the following, an overview of the most important directions of DM candidates is given.

- **Standard Model neutrinos**: Being colour and electrically neutral, neutrinos are the only Standard Model candidates for DM. Although they would correspond to *hot* DM, which would lead to conflicts with current insights on large-scale structure formation, as discussed above, their resulting relic density is examined exemplarily in the following. It is given by:

$$\Omega_\nu h^2 = \sum_{i=1}^{3} \frac{m_i}{93\,\text{eV}} \tag{3.7}$$

where the sum goes over the three generations. The best upper bound on neutrino masses from β-decay experiments states $m_\nu < 2.05$ eV (95% C.L.) [16].[3] Although this bound was stricly-speaking only obtained for electron neutrinos, it can be safely applied to all flavours in this context, since neutrino flavour oscillations constrain the mass differences to be very small. Given the known production rates, this upper mass bound results in a relic density of

$$\Omega_\nu h^2 \lesssim 0.07, \tag{3.8}$$

meaning that neutrinos are by far not sufficient to account for DM. Furthermore, neutrinos were relativistic during structure formation and their free streaming length would be large enough to significantly alter the structure formation history. Neutrino DM would require large structures to build up before small structures, which is strongly disfavoured e.g. by the observation of very old galaxies that must have formed well before larger structures evolved.

- **Sterile neutrinos**: Sterile neutrinos, proposed as having no Standard Model interactions apart from mixing with normal neutrinos, might be significantly heavier than neutrinos and are viable DM candidates [17]. While their detection might be possible via oscillation measurements it is very challenging [18].
- **Axions**: Axions were first proposed as a solution to the *strong CP problem* (see Sect. 2.5). They are expected to be extremely light and only very weakly interacting, meaning that they would not have been in thermal equilibrium in the early universe. It is challenging but possible to identify a range in which axions would be a viable candidate for DM [19].
- **Supersymmetric candidates**: Within R-parity conserving Supersymmetry, the lightest supersymmetric particle is stable and can be a DM candidate. Supersymmetry is discussed in more detail in Chap. 4.
- **Kaluza-Klein states**: The lowest excitation of a Kaluza-Klein tower of states in an assumed extra dimension can act as DM candidate [20].
- **Superheavy Dark Matter**: So-called *WIMPzillas* present an interesting alternative to low- and intermediate-mass DM scenarios. With masses of $10^{12}-10^{16}$ GeV, they would not have followed a thermal evolution. Additionally, they offer a possible solution to the observed cosmic rays with extremely high energies [21].

3.5 Searches for Dark Matter

3.5.1 Direct Detection

If a local relic density of DM is assumed, DM particles χ are expected to traverse the earth. Under the assumption of some weak interaction between χ and Standard

---

[3] Mass bounds from the analysis of CMB anisotropies are even more stringent, but might be evaded under certain circumstances.

Model nucleons, elastic scattering of DM and atomic nuclei occurs. Direct detection (DD) experiments aim at detecting the nuclear recoil of such elastic scatterings. Given the estimated average velocity of DM particles of around 230 km/s, the recoil energy is of the order of 10 keV. The expected event rate is given by:

$$R \propto N \frac{\rho_\chi}{m_\chi} \langle \sigma_{\chi N} \rangle, \qquad (3.9)$$

where N stands for the number of target nuclei, ρ_χ and m_χ denote the DM density and mass, respectively, and $\langle \sigma_{\chi N} \rangle$ gives the average χ-nucleon scattering cross section. The event rate is estimated to be below one event per year and per kg of detection material. Hence, the detectors should comprise a large amount of material and the material should have a high atomic mass in order to increase the expected event rate. Furthermore, the threshold above which the signals of nuclear recoil can be detected needs to be very low ($\mathcal{O}(\text{keV})$), and the expected backgrounds very small and well controlled.

The main background to such experiments comes from photons, muons or electrons scattering with the electrons in the atomic shell ("electronic recoil"), and from neutrons interacting with the nucleus, also leading to a nuclear recoil. Therefore, detectors have to be built in environments where such backgrounds are reduced. Typically, this is ensured by laboratories deep underground and by well shielded detectors. Also, specific muon or photon vetoes can be put in place. Signal and background can be discriminated by measuring the energy loss of the scattering particle via scintillation pulse shapes, charge-to-light ratios or ionisation yields. The actual signal can be detected in various ways and several detector concepts are realised:

- Scintillating crystals (e.g. NaI) can be combined with photomultipliers (example: DAMA/LIBRA [22]).
- Cryogenic detectors measure charge and heat in crystals like Germanium or Silicon (examples: CDMS [23], CRESST [24]). This approach is especially powerful to study lower DM masses.
- A volume of liquid noble gas (Xe, Ar) is combined with photomultipliers and, in most cases, with a Time Projection Chamber (TPC) (examples: LUX [25], Xenon100 [26]). Such detectors dominate, especially at high DM masses.

There are two frontiers that are tried to be pushed. Towards lower DM masses, the nuclear recoil that needs to be detected gets smaller. In order to reduce the energy threshold for a detection further, advanced and probably new detector concepts are needed. Among options discussed in the literature there are detectors involving liquid Helium, DM scattering off cooper pairs of electrons, the usage of nuclear emulsions or DNA detectors. The other direction, mostly pursued via enlarging liquid noble gas detectors, is to lower the limit on the scattering cross section for DM in the typical WIMP mass range of 10–1000 GeV. At scattering cross sections of about four orders of magnitude below what is probed today, scattering of atmospheric neutrinos with the nuclei is expected to make it impossible to detect DM signals below that threshold. A review of DD experiments and their results can be found in

3.5 Searches for Dark Matter

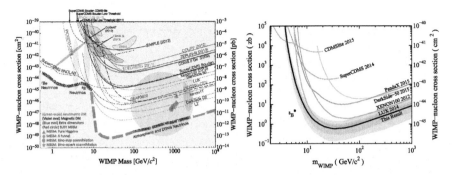

Fig. 3.5 Comparison of DM-nucleon cross section limits (spin-independent), shown as solid lines. Projections are indicated by dotted and dashed lines, possible hints for signals by shaded contours. Interesting regions for several models are shown as shaded regions, the band, where coherent neutrino scattering is expected to limit the sensitivity of DD searches is given in yellow [29]. The current best bound comes from LUX [28], published after the summary was conducted and hence shown separately in the right-hand-side panel

Ref. [27]. Recent limits on the DM-nucleon cross section are summarised in Fig. 3.5. The strongest bounds come from the LUX experiment, reaching down to 0.6 zb at a DM mass of 33 GeV [28]. In the low-mass regime, CDMS-lite provides the strongest constraints for DM masses of 1.6–5.5 GeV [23]. Projected sensitivities of planned experiments start to reach the regime in which neutrino coherent scattering is expected to make a direct detection of DM presumably very challenging. Bounds on chirality suppressed spin-dependent interactions are generally weaker by about five orders of magnitude.

3.5.2 Indirect Detection

Indirect searches for DM try to detect Standard Model annihilation products of DM particles by measuring photons, neutrinos, antimatter, and other objects, especially from regions in the universe, where the DM density is expected to be high and its particles are assumed to move slowly enough to allow for annihilations (e.g. galaxy centre, centre of the sun). Annihilation cross sections consistent with the measured DM relic density are probed by experiments today over a wide mass range. Depending on the wave length considered, different space-based and ground-based telescopes look for DM annihilation products in gamma-ray signals, for example the Fermi LAT [31] (space-based, gamma-ray) and H.E.S.S. [32] (ground-based, gamma-rays) telescopes. It is also possible to detect DM via neutrinos (IceCube [33], ANTARES [34]) or cosmic rays (AMS [35], DAMPE [36]). Figure 3.6 compares current bounds on the DM annihilation cross section from ID experiments. The bounds are obtained for annihilations into muons, b-quarks or W bosons. The strongest bounds come from the scenario of annihilations into b-quarks. The γ-ray measure-

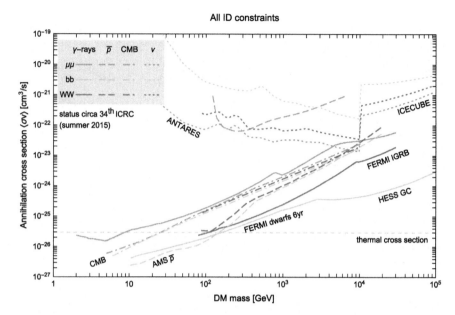

Fig. 3.6 Comparison of bounds on the DM annihilation cross section obtained in different channels, for selected benchmark decays, and from different indirect searches. See Ref. [30] for details

ments (H.E.S.S., Fermi LAT) dominate over constraints from the observed antiproton fraction (AMS) above DM masses of about 150 GeV. The bounds obtained for the W annihilation scenario are slightly weaker, especially for DM masses above 3 TeV. For the annihilation into muons, the strongest constraints come from CMB measurements. The limits obtained by neutrino telescopes (IceCube, ANTARES) are generally weaker, but become important at very high DM masses (above 5 TeV) in all channels. A review of ID experiments and their results is given in Ref. [37].

3.5.3 Dark Matter at the LHC

If some interaction between the Standard Model and the Dark Sector is assumed, DM particles might not only annihilate into Standard Model particles or scatter with nuclei, but could also be produced in particle collisions, for example at the LHC, and detected by the experiments.

Since the DM particles are assumed to interact only weakly with ordinary matter, they would not interact sufficiently strong with the detector material to induce any signal. Hence, they are expected to leave the particle detectors unseen. Their detection requires the presence of an additional, visible object recoiling against the invisible

3.5 Searches for Dark Matter

DM particles. In this way, the momentum imbalance in the transverse plane, resulting from the non-detection of the DM particles, can be used as a discriminating variable.[4]

Hence, many LHC searches for DM target events with such a relatively simple final state of missing transverse momentum and a single, energetic object like a jet, a photon or a vector boson, emitted from the initial state of the *pp* collisions[5]:

$$pp \to \chi + \overline{\chi} + \text{jet}/\gamma/\text{W}/\text{Z}/\ldots \quad (3.10)$$

Here, χ indicates the DM particle. To keep the interpretation of such searches as general as possible, effective field theories [38–42] are often used, as discussed in the next section.

Also searches designed to test more complex, complete theories such as Supersymmetry (introduced in Chap. 4) can be seen as DM searches. If the models provide a DM candidate, constraints on the model then often also constrain the DM parameters. Furthermore, the interplay between searches involving missing transverse momentum, as outlined above, and direct resonance searches can be exploited at the LHC: if it is assumed that the interaction between DM and Standard Model particles happens via a mediating particle, this mediator could not only be produced by SM objects but would also decay back into them, in addition to possible decays into DM. Searches for di-jet resonances for example can be very constraining in certain scenarios.

3.6 Effective Field Theory Description of Dark Matter

An effective field theory (EFT) approach aims at characterising the main features of a physical process, only considering the degrees of freedom which are accessible at a given energy scale. EFT techniques are successfully applied in many areas of physics. As will be discussed below, they also allow for a simple comparison between the collider results and the bounds from direct and indirect DM searches.

The idea of an EFT is that the scale at which some new physics appears is much higher than the one that is probed experimentally. This allows for a simplification of the considered interactions. In the case of one mediator that couples the Standard Model sector to the DM particles, this means that the mediator is far from being possibly produced on-shell.

Considering a simple scenario of such a mediator interacting with both Standard Model and Dark Sector particles, the DM production cross section can be approximated in the following way:

$$\sigma(pp \to \chi\chi) \propto \frac{g_q^2 g_\chi^2}{(Q_{\text{tr}}^2 - M_{\text{med}}^2)^2 + \Gamma_{\text{med}}^2 M_{\text{med}}^2} \simeq -\frac{g_q^2 g_\chi^2}{M_{\text{med}}^4}. \quad (3.11)$$

---

[4]The concept of missing transverse momentum is introduced later in detail in Chap. 5.
[5]In some cases, an emittance from intermediate states might also be possible.

Here, χ stands for the DM particle, M_{med} denotes the mass of the mediator and Γ_{med} its width. Q_{tr} is the momentum transferred in the collision. The couplings of the mediator to the Standard Model sector and the DM particles are given by g_q and g_χ, respectively. To arrive at the approximation made in the last step of Eq. 3.11, the propagator of the mediating particle is expanded in powers of $Q_{\text{tr}}^2/M_{\text{med}}^2$:

$$\frac{g_q g_\chi}{Q_{\text{tr}}^2 - M_{\text{med}}^2} = -\frac{g_q g_\chi}{M_{\text{med}}^2}\left(1 + \frac{Q_{\text{tr}}^2}{M_{\text{med}}^2} + \mathcal{O}\left(\left(\frac{Q_{\text{tr}}^2}{M_{\text{med}}^2}\right)^2\right)\right). \quad (3.12)$$

In the above expression, the width of the mediator is neglected, which is justified in a regime far away from a resonance, such as is the case for an EFT scenario with a very heavy mediator that is far from being possibly produced on-shell. Adapting an EFT approach now means neglecting all contributions apart from the lowest-dimensional term in the expansion of the mediator propagator (Eq. 3.12). As a consequence, EFTs are no longer renormalisible. It is clear that this truncation is only justified as long as $Q_{\text{tr}}^2/M_{\text{med}}^2$ stays small. The so-called cut-off scale of the EFT, Λ, is defined as follows[6]:

$$\Lambda \equiv \frac{M_{\text{med}}}{g_q g_\chi}. \quad (3.13)$$

It is useful to parametrise the cross section and generally presents the only free parameter of the EFT, in addition to the DM mass.

Without loss of generality, a list of effective operators, describing the interaction between Standard Model quarks or gluons and DM can be defined [41]. They are listed in Table 3.1.

The choice of the operator is important, its form determines the interplay between the momentum transfer and the kinematics of the DM particles. Since the measured missing transverse momentum relates directly to the momenta of the DM particles as well as the transverse momentum of the balancing jet, the signal acceptance and detection efficiency strongly depend on the assumed interaction operator.

While the chirality of the operators strongly affects the results from DD, only the effect on the cross section is relevant at the LHC; the kinematics are basically unchanged. This allows to group the fermionic operators into four sets with distinct kinematic characteristics, namely the (pseudo-)scalar, the (axial-)vector and the tensor operators as well as those operators that involve couplings to gluons. By considering one representative operator of each group in an analysis, the full parameter space can be covered, since it is then possible to extend the obtained results to the rest of the operators via rescaling of the cross sections. Analogously, three groups of operators for scalar DM are formed.

In addition to the "standard" scalar operators $D1$–$D4$, the "primed" operators $D1'$–$D4'$ are defined in Table 3.1. They are identical to $D1$–$D4$, but have a different coefficient, independent of the masses of the involved quarks. While a normalisation

---

[6]For some types of interaction this equivalence takes a more complicated form. This is not considered here.

3.6 Effective Field Theory Description of Dark Matter

Table 3.1 List of considered operators. The nomenclature is mostly taken from Ref. [41]. *D* stands for Dirac fermion DM, *C* for complex scalar DM (Real scalar DM operators can be defined as well, but are not considered in the following. No significant phenomenological differences are expected with respect to the complex scalar operators)

Name	Operator	Coefficient
D1	$\bar{\chi}\chi\,\bar{q}q$	m_q/Λ^3
D1′	$\bar{\chi}\chi\,\bar{q}q$	$1/\Lambda^2$
D2	$\bar{\chi}\gamma^5\chi\,\bar{q}q$	im_q/Λ^3
D2′	$\bar{\chi}\gamma^5\chi\,\bar{q}q$	i/Λ^2
D3	$\bar{\chi}\chi\,\bar{q}\gamma^5 q$	im_q/Λ^3
D3′	$\bar{\chi}\chi\,\bar{q}\gamma^5 q$	i/Λ^2
D4	$\bar{\chi}\gamma^5\chi\,\bar{q}\gamma^5 q$	m_q/Λ^3
D4′	$\bar{\chi}\gamma^5\chi\,\bar{q}\gamma^5 q$	$1/\Lambda^2$
D5	$\bar{\chi}\gamma_\mu\chi\,\bar{q}\gamma^\mu q$	$1/\Lambda^2$
D6	$\bar{\chi}\gamma_\mu\gamma^5\chi\,\bar{q}\gamma^\mu q$	$1/\Lambda^2$
D7	$\bar{\chi}\gamma_\mu\chi\,\bar{q}\gamma^\mu\gamma^5 q$	$1/\Lambda^2$
D8	$\bar{\chi}\gamma_\mu\gamma^5\chi\,\bar{q}\gamma^\mu\gamma^5 q$	$1/\Lambda^2$
D9	$\bar{\chi}\sigma_{\mu\nu}\chi\,\bar{q}\sigma^{\mu\nu}q$	$1/\Lambda^2$
D10	$\bar{\chi}\sigma_{\mu\nu}\gamma^5\chi\,\bar{q}\sigma^{\mu\nu}q$	i/Λ^2
D11	$\bar{\chi}\chi\,G^{\mu\nu}G_{\mu\nu}$	$\alpha_s/4\Lambda^3$
D12	$\bar{\chi}\gamma^5\chi\,G^{\mu\nu}G_{\mu\nu}$	$i\alpha_s/4\Lambda^3$
D13	$\bar{\chi}\chi\,G^{\mu\nu}\tilde{G}_{\mu\nu}$	$i\alpha_s/4\Lambda^3$
D14	$\bar{\chi}\gamma^5\chi\,G^{\mu\nu}\tilde{G}_{\mu\nu}$	$\alpha_s/4\Lambda^3$
C1	$\chi^\dagger\chi\,\bar{q}q$	m_q/Λ^2
C2	$\chi^\dagger\chi\,\bar{q}\gamma^5 q$	im_q/Λ^2
C3	$\chi^\dagger\partial_\mu\chi\,\bar{q}\gamma^\mu q$	$1/\Lambda^2$
C4	$\chi^\dagger\partial_\mu\chi\,\bar{q}\gamma^\mu\gamma^5 q$	$1/\Lambda^2$
C5	$\chi^\dagger\chi\,G_{\mu\nu}G^{\mu\nu}$	α_s/Λ^2
C6	$\chi^\dagger\chi\,G_{\mu\nu}\tilde{G}^{\mu\nu}$	$i\alpha_s/\Lambda^2$

proportional to the quark mass is common in many models, motivated by flavour physics and when a mixing between the (pseudo-)scalar mediator and the Standard Model Higgs is assumed, the normalisation can generally have a different form. The primed operators are motivated by integrating out heavy scalars which do not take a vacuum expectation value and therefore do not give rise to quark masses. The unprimed operators $D1$–$D4$ are related to the primed operators $D1'$–$D4'$ by a rescaling:

$$\frac{\mathrm{d}^2\hat{\sigma}}{\mathrm{d}p_\mathrm{T}\mathrm{d}\eta}\bigg|_{D1,D2,D3,D4} = \left(\frac{m_q}{\Lambda}\right)^2 \frac{\mathrm{d}^2\hat{\sigma}}{\mathrm{d}p_\mathrm{T}\mathrm{d}\eta}\bigg|_{D1',D2',D3',D4'}. \qquad (3.14)$$

The choice of DM mass determines the observed kinematics directly, as well as the cross section: since the DM particles are produced on-shell and in pairs, the momentum transfer of the collision has to exceed $2m_\chi$. Assuming higher m_χ hence requires higher momentum transfers, leading to harder spectra in jet momenta and missing transverse momentum. It also means that heavy DM particles (above several TeV) cannot be probed efficiently anymore. However, the change in kinematics is only important for DM masses well above 500 GeV: below, χ can be essentially treated as massless, since the average momentum transfer is generally sufficient to produce the DM particles in the probed region of phase-space. This allows to extrapolate obtained results down to even lower DM masses than explicitly considered in the analyses.

The remaining parameter of the theory, the cut-off scale Λ, has no influence on the resulting spectra but directly determines the signal cross section. The obtained EFT limits are hence usually presented as limits on the cut-off scale.

The key assumption, namely that physics beyond the cut-off scale cannot be probed directly, can be violated. At LHC energies, the momentum transfer in pp collisions at a centre-of-mass energy of several TeV can easily exceed typical bounds on Λ. This issue of limited validity of the LHC approach will be discussed in Chap. 6.

3.6.1 Interplay Between Dark Matter Searches

The EFT approach allows to translate the collider limits on the cut-off scale into limits on the DM-nucleon cross section, probed by direct detection experiments. The expectation value of the partonic operator in the nucleon, considering the kinematic properties of the scattering, is calculated to this end. The conversion can be formulated as follows [43, 44]:

$$\sigma_{\chi N}^{D1} = 1.6 \times 10^{-37} \text{cm}^2 \left(\frac{\mu_\chi}{1\,\text{GeV}}\right)^2 \left(\frac{20\,\text{GeV}}{\Lambda}\right)^6, \tag{3.15}$$

$$\sigma_{\chi N}^{D5,C3} = 1.38 \times 10^{-37} \text{cm}^2 \left(\frac{\mu_\chi}{1\,\text{GeV}}\right)^2 \left(\frac{300\,\text{GeV}}{\Lambda}\right)^4, \tag{3.16}$$

$$\sigma_{\chi N}^{D8,D9} = 4.7 \times 10^{-39} \text{cm}^2 \left(\frac{\mu_\chi}{1\,\text{GeV}}\right)^2 \left(\frac{300\,\text{GeV}}{\Lambda}\right)^4, \tag{3.17}$$

$$\sigma_{\chi N}^{D11} = 3.83 \times 10^{-41} \text{cm}^2 \left(\frac{\mu_\chi}{1\,\text{GeV}}\right)^2 \left(\frac{100\,\text{GeV}}{\Lambda}\right)^6, \tag{3.18}$$

$$\sigma_{\chi N}^{C1} = 2.56 \times 10^{-36} \text{cm}^2 \left(\frac{\mu_\chi}{1\,\text{GeV}}\right)^2 \left(\frac{10\,\text{GeV}}{m_{\text{DM}}}\right)^4 \left(\frac{10\,\text{GeV}}{\Lambda}\right)^4, \tag{3.19}$$

$$\sigma_{\chi N}^{C5} = 7.4 \times 10^{-39} \text{cm}^2 \left(\frac{\mu_\chi}{1\,\text{GeV}}\right)^2 \left(\frac{10\,\text{GeV}}{m_{\text{DM}}}\right)^4 \left(\frac{60\,\text{GeV}}{\Lambda}\right)^4. \tag{3.20}$$

Note that only a subset of possible operators is relevant in the limit of low momentum transfer and hence considered here. The parameter μ_χ denotes the reduced mass of the DM-nucleon system. Depending on the type of interaction, the operators correspond either to spin-dependent or to spin-independent scattering.

In a similar way, the bounds on the cut-off scale for vector and axial-vector operators can be interpreted as a limiting DM annihilation cross section–the quantity that is probed by indirect detection experiments:

$$\sigma_V v_{rel} = \frac{1}{16\pi \Lambda^4} \sum_q \sqrt{1 - \frac{m_q^2}{m_\chi^2}} \left(24 \left(2m_\chi^2 + m_q^2\right) + \frac{8m_\chi^4 - 4m_\chi^2 m_q^2 + 5m_q^4}{m_\chi^2 - m_q^2} v_{rel}^2 \right), \quad (3.21)$$

$$\sigma_A v_{rel} = \frac{1}{16\pi \Lambda^4} \sum_q \sqrt{1 - \frac{m_q^2}{m_\chi^2}} \left(24 m_q^2 + \frac{8m_\chi^4 - 22m_\chi^2 m_q^2 + 17m_q^4}{m_\chi^2 - m_q^2} v_{rel}^2 \right), \quad (3.22)$$

where $v_{rel} \sim 0.24$ is the relative velocity of the annihilating DM particles χ.

References

1. F. Zwicky, Die Rotverschiebung von extragalaktischen Nebeln. Helv. Phys. Acta **6**, 110–127 (1933)
2. H.W. Babcock, The rotation of the Andromeda Nebula. Lick Obs. Bull. **19**, 41–51 (1939). https://doi.org/10.5479/ADS/bib/1939LicOB.19.41B
3. V.C. Rubin, W.K. Ford Jr., Rotation of the Andromeda Nebula from a spectroscopic survey of emission regions. Astrophys. J. **159**, 379–403 (1970). https://doi.org/10.1086/150317
4. E. Corbelli, P. Salucci, The extended rotation curve and the dark matter halo of M33. Mon. Not. R. Astron. Soc. **311**, 441–447 (2000). https://doi.org/10.1046/j.1365-8711.2000.03075.x, arxiv:astro-ph/9909252 [astro-ph]
5. NASA, Bullet cluster, 1e 0657-56 (2006), http://chandra.harvard.edu/photo/2006/1e0657/1e0657.jpg
6. R.A. Alpher, R.C. Herman, On the relative abundance of the elements. Phys. Rev. **74**, 1737–1742 (1948), http://link.aps.org/doi/10.1103/PhysRev.74.1737
7. A.A. Penzias, R.W. Wilson, A measurement of excess antenna temperature at 4080-Mc/s. Astrophys. J. **142**, 419–421 (1965). https://doi.org/10.1086/148307
8. Planck Collaboration, P.A.R. Ade et al., Planck 2015 results. XIII. Cosmological parameters. Astron. Astrophys. **594**, A13 (2016). https://doi.org/10.1051/0004-6361/201525830. arXiv:1502.01589 [astro-ph.CO]
9. B.J. Carr, K. Kohri, Y. Sendouda, J. Yokoyama, New cosmological constraints on primordial black holes. Phys. Rev. D **81**, 104019 (2010). https://doi.org/10.1103/PhysRevD.81.104019. arXiv:0912.5297 [astro-ph.CO]
10. J.D. Bekenstein, Relativistic gravitation theory for the MOND paradigm. Phys. Rev. **D70**, 083509 (2004). https://doi.org/10.1103/PhysRevD.70.083509, https://doi.org/10.1103/PhysRevD.71.069901, arXiv:astro-ph/0403694 [astro-ph]
11. S.S. McGaugh, A tale of two paradigms: the mutual incommensurability of ΛCDM and MOND. Can. J. Phys. **93**(2), 250–259 (2015). https://doi.org/10.1139/cjp-2014-0203. arXiv:1404.7525 [astro-ph.CO]
12. C.R. Contaldi, T. Wiseman, B. Withers, TeVeS gets caught on caustics. Phys. Rev. D **78**, 044034 (2008). https://doi.org/10.1103/PhysRevD.78.044034. arXiv:0802.1215 [gr-qc]
13. C. Patrignani, Review of particle physics. Chin. Phys. **C40**(10), 100001 (2016). https://doi.org/10.1088/1674-1137/40/10/100001

14. D. Hooper, Particle dark matter, in *Proceedings of Theoretical Advanced Study Institute in Elementary Particle Physics on The dawn of the LHC era (TASI 2008)*, Boulder, USA, 2–27 June 2008 (2010), pp. 709–764. https://doi.org/10.1142/9789812838360_0014, http://lss.fnal.gov/cgi-bin/find_paper.pl?conf-09-025, arXiv:0901.4090 [hep-ph]
15. Planck Collaboration, P.A.R. Ade et al., Planck 2013 results. XVI. Cosmological parameters. Astron. Astrophys. **571**, A16 (2014). https://doi.org/10.1051/0004-6361/201321591. arXiv:1303.5076 [astro-ph.CO]
16. Troitsk Collaboration, V.N. Aseev et al., An upper limit on electron antineutrino mass from Troitsk experiment. Phys. Rev. **D84**, 112003 (2011). https://doi.org/10.1103/PhysRevD.84.112003, arXiv:1108.5034 [hep-ex]
17. T. Asaka, M. Laine, M. Shaposhnikov, Lightest sterile neutrino abundance within the nuMSM. JHEP **01**, 091 (2007). https://doi.org/10.1088/1126-6708/2007/01/091, https://doi.org/10.1007/JHEP02(2015)028, arXiv:hep-ph/0612182 [hep-ph]
18. Daya Bay Collaboration, F.P. An et al., Search for a light sterile neutrino at Daya Bay. Phys. Rev. Lett. **113**, 141802 (2014). https://doi.org/10.1103/PhysRevLett.113.141802. arXiv:1407.7259 [hep-ex]
19. L.D. Duffy, K. van Bibber, Axions as dark matter particles. New J. Phys. **11**, 105008 (2009). https://doi.org/10.1088/1367-2630/11/10/105008. arXiv:0904.3346 [hep-ph]
20. H.-C. Cheng, J.L. Feng, K.T. Matchev, Kaluza-Klein dark matter. Phys. Rev. Lett. **89**, 211301 (2002). https://doi.org/10.1103/PhysRevLett.89.211301. arXiv:hep-ph/0207125 [hep-ph]
21. E.W. Kolb, D.J.H. Chung, A. Riotto, WIMPzillas!, in *Proceedings, 2nd La Plata Meeting, Trends in Theoretical Physics II*, Buenos Aires, Argentina, 29 Nov–4 Dec 1998, pp. 91–105, http://lss.fnal.gov/cgi-bin/find_paper.pl?conf-98-325, arXiv:hep-ph/9810361 [hep-ph]
22. D.A.M.A. Collaboration, R. Bernabei et al., The DAMA/LIBRA apparatus. Nucl. Instrum. Meth. **A592**, 297–315 (2008). https://doi.org/10.1016/j.nima.2008.04.082. arXiv:0804.2738 [astro-ph]
23. R. SuperCDMS Collaboration, Agnese et al., New results from the search for low-mass weakly interacting massive particles with the CDMS low ionization threshold experiment. Phys. Rev. Lett. **116**(7), 071301 (2016). https://doi.org/10.1103/PhysRevLett.116.071301. arXiv:1509.02448 [astro-ph.CO]
24. CRESST-II Collaboration, G. Angloher et al., Results on low mass WIMPs using an upgraded CRESST-II detector. Eur. Phys. J. **C74**(12), 3184 (2014). https://doi.org/10.1140/epjc/s10052-014-3184-9, arXiv:1407.3146 [astro-ph.CO]
25. LUX Collaboration, D.S. Akerib et al., The large underground xenon (LUX) experiment. Nucl. Instrum. Meth. **A704**, 111–126 (2013). https://doi.org/10.1016/j.nima.2012.11.135, arXiv:1211.3788 [physics.ins-det]
26. XENON100 Collaboration, E. Aprile et al., The XENON100 dark matter experiment. Astropart. Phys. **35**, 573–590 (2012). https://doi.org/10.1016/j.astropartphys.2012.01.003. arXiv:1107.2155 [astro-ph.IM]
27. L. Baudis, Direct dark matter detection: the next decade. Phys. Dark Univ. **1**, 94–108 (2012). https://doi.org/10.1016/j.dark.2012.10.006. arXiv:1211.7222 [astro-ph.IM]
28. L.U.X. Collaboration, D.S. Akerib et al., Improved limits on scattering of weakly interacting massive particles from reanalysis of 2013 LUX data. Phys. Rev. Lett. **116**(16), 161301 (2016). https://doi.org/10.1103/PhysRevLett.116.161301. arXiv:1512.03506 [astro-ph.CO]
29. P. Cushman et al., Working group report: WIMP dark matter direct detection, in *Proceedings, Community Summer Study 2013: Snowmass on the Mississippi (CSS2013)*, Minneapolis, MN, USA, 29 July–6 Aug 2013, https://inspirehep.net/record/1262767/files/arXiv:1310.8327.pdf, arXiv:1310.8327 [hep-ex]
30. M. Cirelli, Status of Indirect (and Direct) Dark Matter searches, arXiv:1511.02031 [astro-ph.HE]
31. M. Fermi-LAT Collaboration, Ackermann et al., Constraining dark matter models from a combined analysis of milky way satellites with the fermi large area telescope. Phys. Rev. Lett. **107**, 241302 (2011). https://doi.org/10.1103/PhysRevLett.107.241302. arXiv:1108.3546 [astro-ph.HE]

References

32. H.E.S.S. Collaboration, H. Abdallah et al., Search for dark matter annihilations towards the inner Galactic halo from 10 years of observations with H.E.S.S. Phys. Rev. Lett. **117**(11), 111301 (2016). https://doi.org/10.1103/PhysRevLett.117.111301. arXiv:1607.08142 [astro-ph.HE]
33. F. Halzen, S.R. Klein, IceCube: an instrument for neutrino astronomy. Rev. Sci. Instrum. **81**, 081101 (2010). https://doi.org/10.1063/1.3480478. arXiv:1007.1247 [astro-ph.HE]
34. ANTARES Collaboration, S. Adrian-Martinez et al., Limits on dark matter annihilation in the sun using the ANTARES neutrino telescope. Phys. Lett. **B759**, 69–74 (2016). https://doi.org/10.1016/j.physletb.2016.05.019, arXiv:1603.02228 [astro-ph.HE]
35. A.M.S. Collaboration, M. Aguilar et al., First result from the alpha magnetic spectrometer on the international space station: precision measurement of the positron fraction in primary cosmic rays of 0.5-350 GeV. Phys. Rev. Lett. **110**, 141102 (2013). https://doi.org/10.1103/PhysRevLett.110.141102
36. C. Jin, Dark matter particle explorer: the first chinese cosmic ray and hard γ-ray detector in space. Chin. J. Space Sci. **34**(5), 550 (2014). https://doi.org/10.11728/cjss2014.05.550, http://www.cjss.ac.cn/CN/abstract/article_2067.shtml
37. M. Cirelli, Indirect searches for dark matter: a status review. Pramana **79**, 1021–1043 (2012). https://doi.org/10.1007/s12043-012-0419-x. arXiv:1202.1454 [hep-ph]
38. M. Beltran, D. Hooper, E.W. Kolb, Z.A.C. Krusberg, T.M.P. Tait, Maverick dark matter at colliders. JHEP **09**, 037 (2010). https://doi.org/10.1007/JHEP09(2010)037. arXiv:1002.4137 [hep-ph]
39. J. Goodman, M. Ibe, A. Rajaraman, W. Shepherd, T.M.P. Tait, H.-B. Yu, Constraints on light majorana dark matter from colliders. Phys. Lett. B **695**, 185–188 (2011). https://doi.org/10.1016/j.physletb.2010.11.009. arXiv:1005.1286 [hep-ph]
40. Y. Bai, P.J. Fox, R. Harnik, The Tevatron at the frontier of dark matter direct detection. JHEP **12**, 048 (2010). https://doi.org/10.1007/JHEP12(2010)048. arXiv:1005.3797 [hep-ph]
41. J. Goodman, M. Ibe, A. Rajaraman, W. Shepherd, T.M.P. Tait, H.-B. Yu, Constraints on dark matter from colliders. Phys. Rev. D **82**, 116010 (2010). https://doi.org/10.1103/PhysRevD.82.116010. arXiv:1008.1783 [hep-ph]
42. P.J. Fox, R. Harnik, J. Kopp, Y. Tsai, Missing energy signatures of dark matter at the LHC. Phys. Rev. D **85**, 056011 (2012). https://doi.org/10.1103/PhysRevD.85.056011. arXiv:1109.4398 [hep-ph]
43. J. Goodman, M. Ibe, A. Rajaraman, W. Shepherd, M. Tait, H. Yu, Constraints on dark matter from colliders. Phys. Rev. D **82**, 116010 (2010). https://doi.org/10.1103/PhysRevD.82.116010. arXiv:hep-ph/1008.1783v2
44. ATLAS Collaboration, Search for dark matter candidates and large extra dimensions in events with a jet and missing transverse momentum with the ATLAS detector. JHEP **1304**, 075 (2013). https://doi.org/10.1007/JHEP04(2013)075, arXiv:1210.4491 [hep-ex]

Chapter 4
Supersymmetry

Supersymmetry (SUSY) is one of the most popular and well-motivated extension of the Standard Model, since it is able to solve many of its problems at once, while being based on a theoretically simple and beautiful idea. Via the introduction of an additional symmetry, SUSY connects bosons and fermions and proposes bosonic (fermionic) superpartners for Standard Model fermions (bosons), with otherwise identical charges and masses. This means that for each fermion loop contributing to the Higgs boson mass, there is now also a boson loop of equal magnitude but opposite sign. In this way, SUSY offers a solution to the technical aspect of the hierarchy problem.

The observed mass of the discovered Higgs Boson however severely challenges naive SUSY models, in which the Higgs mass is predicted to be similar to the mass of the Z boson at tree-level. Loop corrections could lift its mass up to the observed 125 GeV but would require the scalar top quark partner to be much heavier than the top–which in turn affects the cancellation of contributions to the Higgs mass and makes some re-introduction of fine-tuning necessary.

But there are also other motivations for SUSY: the renormalisation group evolution of the three gauge coupling constants of the Standard Model is sensitive to the particle content of the theory. Given the Standard Model, the coupling constants of the different interactions do not "meet" at a common energy scale. However, with the addition of the proposed SUSY particles, the renormalisation group equation predicts them to converge at approximately 10^{16} GeV. This would allow to formulate the Standard Model gauge group within a larger symmetry group $SU(5) \subset SU(3) \times SU(2) \times U(1)$, and possibly to formulate a grand unified theory ("GUT").

Furthermore, SUSY can be connected to general relativity: imposing SUSY as a local symmetry allows to formulate a class of models known as *supergravity*. SUSY is also a necessary prerequisite of string theories and can be connected to cosmological inflation. In addition, it can provide a candidate for Dark Matter.

© Springer International Publishing AG, part of Springer Nature 2018
J. Gramling, *Search for Dark Matter with the ATLAS Detector*,
Springer Theses, https://doi.org/10.1007/978-3-319-95016-7_4

In summary, even if the observed Higgs mass and the direct LHC bounds disfavour simple versions of SUSY, many motivations for this idea exist even beyond natural models of SUSY that provide a solution to the hierarchy problem.

In the following, the main concepts of SUSY are discussed in Sect. 4.1 before the minimal supersymmetric extension of the Standard Model is introduced in Sect. 4.2. Finally, the focus is put on the supersymmetric partner of the top quark, the *stop*, in Sect. 4.3 since a search for stops will be presented in Chap. 9.

4.1 Main Concepts of Supersymmetry

The idea of SUSY is the introduction of an additional symmetry, the so-called *Supersymmetry*, which connects fermions and bosons:

$$Q|\text{fermion}\rangle = |\text{boson}\rangle \quad \text{and} \quad Q|\text{boson}\rangle = |\text{fermion}\rangle . \tag{4.1}$$

Here Q denotes the SUSY generator. The SUSY algebra is a nontrivial extension of the Poincaré algebra that covers spacetime transformations. It circumvents the Coleman–Mandula theorem [1] that states that space-time and internal symmetries cannot be combined in a non-trivial way, by allowing both commuting and anticommuting symmetry generators [2]. The generators have to be (Weyl) spinors: Q_α. The crucial new anticommutator is given by:

$$\{Q_\alpha, \bar{Q}_{\dot{\alpha}}\} = 2(\sigma^\mu)_{\alpha\dot{\alpha}} P_\mu . \tag{4.2}$$

It can be explicitly seen here, that the internal symmetry, SUSY, is related to the space-time Poincaré symmetry, since the momentum operator P_μ appears on the right-hand side. This can be understood as a "mixing" of internal and space-time transformations: while SUSY generators transform bosons into fermions and vice versa, the anticommutator of two such transformations yields a translation in spacetime.

All other anti-commutation relations between the Qs and commutation relations between the Qs and Ps vanish. Since Q commutes with the energy-momentum operator P^μ and its square P^2, SUSY predicts the masses of bosonic (fermonic) particles and their fermionic (bosonic) superpartners to be equal:

$$P^2 Q|b\rangle = P^2|f\rangle = m_f^2|f\rangle , \tag{4.3}$$

and, on the other hand:

$$P^2 Q|b\rangle = Q P^2|b\rangle = m_b^2 Q|b\rangle = m_b^2|f\rangle , \tag{4.4}$$

leading to:

$$m_b^2 = m_f^2 . \tag{4.5}$$

4.2 The Minimal Supersymmetric Standard Model

The minimal theory that extends the Standard Model to a supersymmetric theory is called *Minimal Supersymmetric Standard Model* (MSSM). Each fermion being related to a bosonic partner with the same quantum numbers and vice versa, leads to an enlarged particle content: since no pair of Standard Model particles could be combined in such a supermultiplet, the model is extended by partner particles for each of the known particles in the Standard Model, having a spin that is different by 1/2. The scalar partners are called as the particles, but an *s* is prepended, the fermionic partners are denoted by *ino* appended to the name.

The MSSM is *minimal* in two ways. First, it assumes the minimal gauge group, based on the Standard Model symmetries: to avoid additional gauge interactions that would arise from spin-1 superpartners of Standard Model fermions, fermion partners (*sfermions*) are assumed to be scalar. The Standard Model spin-1 gauge bosons form with their spin-1/2 superpartners, the gauginos (bino $\tilde{B}$, winos $\tilde{W}_{1-3}$ and gluinos $\tilde{G}_{1-8}$) the *vector supermultiplets*.

Second, the MSSM assumes minimal particle content: there are only three generations of spin-1/2 fermions and their partners assumed,[1] as in the Standard Model. The left- and right-handed fermion fields belong to *chiral supermultiplets* together with their spin-0 SUSY partners, the *squarks* and *sleptons*. The matter content of the MSSM is hence formed by three generations of *chiral supermultiplets*.

The Higgs sector differs from the Standard Model structure: the MSSM contains two *chiral supermultiplets* with hypercharges $+1$ and -1 containing two complex Higgs doublets, H_u and H_d, together with their fermionic partners, the *higgsinos*. This is the minimal structure required for renormalisability of the theory, since gauge anomalies arise if the sum of the fermionic hypercharges does not vanish. The scalar components:

$$H_d = \begin{pmatrix} H_d^0 \\ H_d^- \end{pmatrix} \quad H_u = \begin{pmatrix} H_u^+ \\ H_u^0 \end{pmatrix} \tag{4.6}$$

give mass separately to the up-type and down-type fermions.

Their spin-1/2 superpartners, the higgsinos, mix with the gauginos to form the physical mass eigenstates, the *charginos*: $\tilde{\chi}_1^\pm$, $\tilde{\chi}_2^\pm$ and *neutralinos*: $\chi_{1,2,3,4}^0$. The indices are chosen such that they represent the mass order: $m(\tilde{\chi}_1) < m(\tilde{\chi}_2)$.

An overview of the gauge and mass eigenstates of the additional particle content in the MSSM is given in Table 4.1.

4.2.1 Breaking of Supersymmetry

Since supersymmetric partners with masses equal to the Standard Model particles would have been easily discovered, any SUSY model that aims at providing a realistic

---

[1] No right-handed neutrinos are added.

Table 4.1 Overview of SUSY partners of Standard Model particles and their gauge and mass eigenstates as well as the content of the extended Higgs sector [3]

Particle	Spin	R-Parity	Gauge eigenstates	Mass eigenstates
Higgs bosons	0	+1	$H_u^0, H_d^0, H_u^+, H_d^-$	$h^0\ H^0\ A^0\ H^\pm$
Squarks	0	−1	$\tilde{u}_L, \tilde{u}_R, \tilde{d}_L, \tilde{d}_R$	Same
			$\tilde{c}_L, \tilde{c}_R, \tilde{s}_L, \tilde{s}_R$	Same
			$\tilde{t}_L, \tilde{t}_R, \tilde{b}_L, \tilde{b}_R$	$\tilde{t}_1, \tilde{t}_2, \tilde{b}_1, \tilde{b}_2$
Sleptons	0	−1	$\tilde{e}_L, \tilde{e}_R, \tilde{\nu}_e$	Same
			$\tilde{\mu}_L, \tilde{\mu}_R, \tilde{\nu}_\mu$	Same
			$\tilde{\tau}_L, \tilde{\tau}_R, \tilde{\nu}_\tau$	$\tilde{\tau}_1, \tilde{\tau}_2, \tilde{\nu}_\tau$
Neutralinos	1/2	−1	$\tilde{B}^0, \tilde{W}^0, \tilde{H}_u^0, \tilde{H}_d^0$	$\tilde{\chi}_1^0\ \tilde{\chi}_1^0\ \tilde{\chi}_2^0\ \tilde{\chi}_3^0\ \tilde{\chi}_4^0$
Charginos	1/2	−1	$\tilde{W}^\pm, \tilde{H}_u^+, \tilde{H}_d^-$	$\tilde{\chi}_1^\pm\ \tilde{\chi}_2^\pm$
Gluino	1/2	−1	$\tilde{g}$	Same

phenomenology must contain a breaking mechanism of Supersymmetry, manifesting itself as SUSY partners being heavier than their SM counterparts.

To maintain the above-mentioned desirable features of SUSY, theories generally consider spontaneous symmetry breaking (SSB), meaning that the underlying Lagrangian is supersymmetric, but the vacuum state realised in nature is not. SUSY would hence emerge at higher energies.

There are many possibilities to realise such a SUSY SSB, and it is not clear which ones are preferable. However, these proposals generally involve extra particles and interactions at higher scales, allowing to ignore the exact high-energy mechanism by introducing an explicit symmetry-breaking term in the Lagrangian.

As mentioned above, unbroken SUSY allows to cancel the correction terms to scalar masses such as the Higgs mass exactly: first, by requiring equal masses of superpartners, and second, by the relation of the scalar and fermionic couplings:

$$\lambda_s = |\lambda_f|^2. \tag{4.7}$$

For a broken symmetry, to avoid reintroducing quadratic divergences, the above coupling relation should be maintained. Therefore, only mass terms and couplings of positive mass dimension, called *soft terms*, are proposed to enter the SUSY-breaking Lagrangian, allowing to write:

$$\mathcal{L} = \mathcal{L}_{SUSY} + \mathcal{L}_{soft}. \tag{4.8}$$

$\mathcal{L}_{soft}$ then contains mass terms for gauginos and sfermions, mass and bilinear terms for the Higgs sector and trilinear couplings between sfermions and the Higgs sector.

The presence of additional light particles associated with the SUSY breaking sector would clearly be problematic, since such states are not observed. This can be avoided if the symmetry-breaking sector is taken to be *hidden*, i.e. taken to only interact with the SM and SUSY sectors via a messenger. If this messenger is heavy, the extra sector and all the particles it contains is effectively hidden from observation.

4.2.2 R-Parity

The MSSM assumes an additional $U(1)$ symmetry that leads to a multiplicative quantum number, called *R-parity*, which is defined as follows:

$$R = (-1)^{3(B-L)+2S} . \qquad (4.9)$$

B is the baryon number, L the lepton number, S denotes the spin. Following this definition, the R-parity distinguishes particles ($R = +1$) and SUSY partners ($R = -1$).

Conservation of R-parity is motivated in order to avoid lepton and baryon number violating terms in the MSSM. Generally, R-parity conservation leads to the following important phenomenological constraints:

- Supersymmetric particles are produced in pairs, such that R-parity "cancels".
- In decays of supersymmetric particles there always needs to be an odd number of SUSY particles in the final state. Hence, the lightest supersymmetric particle (LSP) cannot decay and is stable.

If R-parity is assumed and the lightest SUSY particle is given by the lightest neutralino, $\tilde{\chi}_1^0$, it can be an excellent candidate for Dark Matter, since it is neutral and stable. Beyond the MSSM, there are several models allowing for violation of R-parity conservation. They lead to distinct phenomenologies with less or no missing transverse momentum and possibly displaced vertices in the final state [4].

4.2.3 Reducing Parameters

The MSSM as presented above adds 105 free parameters to the 19 parameters of the SM. The most relevant ones for the following discussion are:

- μ: Higgs mass parameter with $\mathcal{L} \supset \mu H_u H_d$. Generally, μ is taken to be complex. However, its phase possibly introduces large CP-violating terms and hence μ is often assumed to be real.
- $\mathbf{M_i}$: mass parameters for the gauginos, appearing in $\mathcal{L}_{soft}$.
- $\tan \beta$: ratio of vacuum expectation values of the two Higgs doublets, $\tan \beta = v_u/v_d$.
- $\mathbf{m_{\tilde{Q}_i}}, \mathbf{m_{\tilde{u}_i}}, \mathbf{m_{\tilde{d}_i}}$: masses of left-handed and right-handed squarks, appearing in $\mathcal{L}_{soft}$.
- $\mathbf{m_{\tilde{L}_i}}, \mathbf{m_{\tilde{e}_i}}$: masses of left-handed and right-handed sleptons, appearing in $\mathcal{L}_{soft}$.

Clearly, the predictive power of a model with so many free parameters is limited. It is interesting to note here that explicitly the quest for naturalness led to propose a model which might be conceptually simple and aesthetic, but is by far not simple regarding new entities and parameters that are introduced. The model-building guidelines of naturalness and simplicity seem to be in conflict. In order to reduce the number of free parameters, several scenarios can be followed.

4.2.3.1 Assumption of a SUSY-Breaking Scenario

- **mSUGRA**: in so-called *minimal Supergravity*, the mediation between the SUSY-breaking sector and the MSSM is taken to happen via gravitational interaction. Models of mSUGRA generate the soft SUSY-breaking terms via the supersymmetric equivalent of the Higgs mechanism. The idea of mSUGRA is to fix certain parameters at the GUT scale and use the renormalisation group equation to obtain their values at the relevant scale. To this end, it assumes gauge coupling unification, the unification of gaugino masses, universal scalar (sfermion and Higgs) masses and universal trilinear couplings at the GUT scale.

  In this setting, the theory requires only four parameters, normally chosen to be: m_0, the common mass of the scalars, $m_{1/2}$, the common mass of the gauginos and higgsinos, A_0, the universal trilinear coupling, and $\tan \beta$. Additionally, the sign of μ needs to be assumed.

  Anomaly-mediated symmetry breaking (**AMSB**) presents a special case of gravity mediation, in which the mediation is formulated as a conformal anomaly.

- **GMSB**: in gauge mediated symmetry breaking, the mediation takes place through the Standard Model's gauge interactions. Typically, a hidden sector breaks SUSY and communicates to massive messenger fields that are charged under the Standard Model. These messenger fields induce a gaugino mass via one-loop diagrams which is then transmitted to the scalar superpartners via two-loop processes. With the Higgs boson being discovered at 125 GeV, *stops* with high masses above 2 TeV are required in this scenario.

4.2.3.2 Phenomenological MSSM

By imposing empirically motivated assumptions, the so-called *phenomenological MSSM*
(pMSSM) can be constructed. The assumptions are the following:

- **No new source of CP violation**: additional CP violation is constrained in particular by measurements of the electron and neutron electric moments. Eliminating all phases from the MSSM prohibits any new source of CP violation.
- **No flavour-changing neutral currents**: The non-diagonal terms in the sfermion mass matrices and in the trilinear coupling matrices can induce significant flavour-changing neutral currents (FCNCs) which are severely constrained by present experimental data. This can be circumvented by assuming *flavour universality* (sfermions have very similar masses) and *flavour alignment* (the mass matrices of quarks and squarks are almost proportional to each other). It is commonly assumed that both the matrices for the sfermion masses and for the trilinear couplings are diagonal.
- **First and second generation universality**: while there are no constraints on the third generation masses, experimental data, especially from neutral Kaon mixing,

4.2 The Minimal Supersymmetric Standard Model

severely limit the mass splitting between the first- and second-generation squarks.[2] One further assumes that the trilinear couplings are the same for these two generations. Since they are generally proportional to the fermion masses, these couplings are negligible for the first and second generation and can be set to zero. They do become important for the third generation.

These restrictions reduce the number of parameters to 19:

- **Higgs sector**: $\tan\beta$, M_A, the mass of pseudoscalar Higgs, μ, the higgsino mass
- **Gaugino sector**: M_1, M_2, M_3, the masses of Bino, Wino and Gluino
- **Squark masses**: $m_{\tilde{q}}$, $m_{\tilde{u}_R}$, $m_{\tilde{d}_R}$, the masses for the (degenerate) first and second generation, and $m_{\tilde{Q}}$, $m_{\tilde{t}_R}$, $m_{\tilde{b}_R}$, the masses for the third generation
- **Slepton masses**: $m_{\tilde{l}}$, $m_{\tilde{e}_R}$, the masses for the (degenerate) first and second generation, and $m_{\tilde{L}}$, $m_{\tilde{\tau}_R}$, the masses for the third generation
- **The third generation trilinear couplings**: A_t, A_b, A_τ.

4.2.3.3 Simplified Models

Another approach to reduce the complexity of SUSY models and to allow for experimental tests of certain aspects of a model is the construction of so-called *Simplified Models*. Such models make the following assumptions:

- They are constructed such that only one decay mode is considered at a time, assuming it to have a branching ratio of 100%. This also means that interferences between different decay modes are neglected.
- All sparticles not involved in the decay are assumed to be *decoupled*, i.e. having much higher masses.
- The masses of the involved sparticles are generally treated as free parameters.

Simplified Models of top squark production will be used for the interpretation of the results from the search for new physics in final states with top quarks and missing transverse momentum presented in Chap. 9.

4.3 Supersymmetric Top Quark Partners

The SUSY partners of the top quark play a special role: first off, since the top is the heaviest fermion in the Standard Model, it gives the largest contribution to the problematic Higgs mass correction terms. Hence, it is the most crucial point at which solutions to the hierarchy problem should show up. Second, the large possible mixing allows for one of the stops to be light without severe phenomenological consequences such as large FCNCs. Consequently, stop searches are an excellent place to look for SUSY at the LHC.

---

[2] This is the case unless squarks are taken to be significantly heavier than 1 TeV.

Each of the SUSY partners contributes to the Higgs mass correction δm_H^2. The terms are of equal magnitude as the contributions from the Standard Model fermions, but with opposite sign and hence cancel the quadratic dependence on the cut-off scale. If SUSY masses were equal to Standard Model masses, the cancellation would be exact. Otherwise, it is reduced to a logarithmic divergence:

$$\delta m_H^2 \propto \frac{\lambda_f^2 N_c^f}{8\pi^2}(m_{\tilde{f}}^2 - m_f^2)\ln(\Lambda^2/m_{\tilde{f}}^2). \tag{4.10}$$

The larger the difference in mass between the top and the stop, the larger the contribution to the Higgs mass that remains and the larger the still required amount of fine-tuning. Naturalness arguments hence prefer a light stop, close to the top mass.

Since the Standard Model fermions are chiral spinors and the number of degrees of freedom between the fermionic and the bosonic sector needs to be identical, each fermion has two scalar partners, one is the partner of the left-handed fermion, $\tilde{f}_L$, and one of the right-handed fermion, $\tilde{f}_R$. In general, these two do not have to be mass eigenstates and can mix. Since the mixing of these states is proportional to the mass of their fermion partner, the mixing is taken to be small, except for the third generation. Especially the possibly large mixing of partner states of the top quark can be expected to lead to significant mass splittings between the mass eigenstates $\tilde{t}_1$ and $\tilde{t}_2$, where the lighter one, $\tilde{t}_1$, is therefore considerably lighter than all other squarks.

Overall, there are the following contributions to the mass of the stops:

- The squared-mass terms proportional to $\tilde{t}_L^* \tilde{t}_L$ and $\tilde{t}_R^* \tilde{t}_R$ are given by $m_{Q_3}^2 + (1/2 - 2/3\sin^2\theta_W)\cos(2\beta)m_Z^2$ and $m_{\tilde{u}_3}^2 + (2/3\sin^2\theta_W)\cos(2\beta)m_Z^2$ respectively. These terms occur analogously for the other generations.
- Contributions equal to m_t stem from $y_t^2 H_u^{0*} H_u^0 \tilde{t}_L^* \tilde{t}_L$ and $y_t^2 H_u^{0*} H_u^0 \tilde{t}_R^* \tilde{t}_R$, where the Higgs field gets replaced by its vev. For the other generations, these contributions exist as well, but are unimportant due to the small Yukawa coupling.
- Expressions related to the so-called F-Terms take the form: $-\mu^* y_t \tilde{\tilde{t}} \tilde{t} H_d^{0*} + c.c.$. They read $-\mu^* v y_t \cos\beta \tilde{t}_R^* \tilde{t}_L + c.c.$ once H_d^0 gets replaced by its vev.
- Contribution from soft trilinear couplings $a_t \tilde{\bar{t}} \tilde{Q}_3 H_u^0 + c.c.$ become $a_t v \sin\beta \tilde{t}_L \tilde{t}_R^* + c.c.$.

One can now define a mass matrix, containing these different contributions to the stop masses:

$$\mathbf{M}_{\tilde{t}}^2 = \begin{pmatrix} m_{Q_3}^2 + m_t^2 + (1/2 - 2/3\sin^2\theta_W)\cos(2\beta)m_Z^2 & -m_t(A_t + \mu\cot\beta) \\ -m_t(A_t + \mu\cot\beta) & m_{\tilde{u}_3}^2 + m_t^2 + 2/3\sin^2\theta_W\cos(2\beta)m_Z^2 \end{pmatrix}. \tag{4.11}$$

Then, the terms relevant to the stop mass in the Lagrangian can be written as:

$$\mathcal{L}_{m_{\tilde{t}}} = -\begin{pmatrix} \tilde{t}_L^* & \tilde{t}_R^* \end{pmatrix} \mathbf{M}_{\tilde{t}}^2 \begin{pmatrix} \tilde{t}_L \\ \tilde{t}_R \end{pmatrix}. \tag{4.12}$$

4.3 Supersymmetric Top Quark Partners

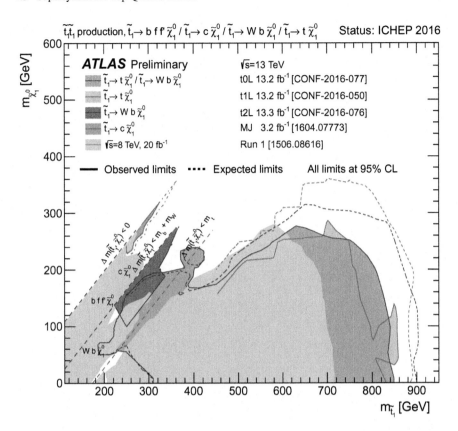

Fig. 4.1 Summary results from ATLAS searches for stop pair production based on 13.2 fb^{-1} of pp collision data taken at $\sqrt{s} = 13$ TeV. Exclusion limits at 95% CL are shown in the $\tilde{t}_1$–$\tilde{\chi}_1^0$ mass plane. The dashed and solid lines show the expected and observed limits, respectively, including all uncertainties except the theoretical signal cross section uncertainty. Four decay modes are considered separately with 100% $\mathcal{BR}$: $\tilde{t}_1 \to t + \tilde{\chi}_1^0$ (where the $\tilde{t}_1$ is mostly right-handed), $\tilde{t}_1 \to W + b + \tilde{\chi}_1^0$ (3-body decay for $m(\tilde{t}_1) < m(t) + m(\tilde{\chi}_1^0)$), $\tilde{t}_1 \to c + \tilde{\chi}_1^0$ and $\tilde{t}_1 \to f\bar{f} + b + \tilde{\chi}_1^0$ (4-body decay). The latter two decay modes are superimposed. Figure from Ref. [5]

Current experimental bounds on the stop mass reach up to $m(\tilde{t}_1) = 850$ GeV for light neutralinos. A summary of the results from ATLAS stop searches is given in Fig. 4.1, where several channels are considered for different mass splittings between stop and neutralino, leading to different decay scenarios. The result obtained in the one-lepton channel is discussed in detail in Chap. 9.

References

1. S.R. Coleman, J. Mandula, All possible symmetries of the S matrix. Phys. Rev. **159**, 1251–1256 (1967). https://doi.org/10.1103/PhysRev.159.1251
2. R. Haag, J.T. Lopuszanski, M. Sohnius, All possible generators of supersymmetries of the S matrix. Nucl. Phys. B **88**, 257 (1975). https://doi.org/10.1016/0550-3213(75)90279-5
3. S.P. Martin, A supersymmetry primer, arXiv:hep-ph/9709356 [hep-ph]
4. R. Barbier et al., R-parity violating supersymmetry. Phys. Rept. **420**, 1–202 (2005). https://doi.org/10.1016/j.physrep.2005.08.006, arXiv:hep-ph/0406039 [hep-ph]
5. ATLAS Collaboration, Summary of top squark searches (2016), https://atlas.web.cern.ch/Atlas/GROUPS/PHYSICS/CombinedSummaryPlots/SUSY/ATLAS_SUSY_Stop_tLSPATLAS_SUSY_Stop_tLSP.pdf

Chapter 5
The ATLAS Detector at the LHC

In order to learn more about fundamental interactions, particle physics offers three directions: increasing the energy, increasing the intensity or increasing the precision. The Large Hadron Collider (LHC) clearly pushes the energy frontier by achieving unprecedented collision energies. But it also provides very large datasets to test very weak interactions and the excellent performance of its experiments allows to improve the precision of measurements on some parameters and properties significantly. Being a hadron collider, the momentum transfer in the collisions is not fixed to one exact energy (as opposed to electron-positron colliders) and it is an excellent machine for discovering new particles.

With the increase of the collision energy, not only the accelerator has to grow,[1] also the detectors need to be optimised for higher-energetic particles and hence become larger. With the size of the detector and the increase in complexity also the experimental collaborations reached unprecedented sizes at the LHC: ATLAS for example counts more than 5000 [1] collaborators, that operate the experiment, including the analysis of the data. In the following, an overview of the ATLAS experiment at the LHC is given. A focus is put on the trigger system since I contributed to this area in several ways: I was responsible for the maintenance and development of the monitoring of the electron-photon trigger and implemented a tag-and-probe algorithm for efficiency measurements to be run at the first stage of reconstruction. Later, I contributed to the software validation effort for the jet trigger. Between 2014 and 2016 I was actively involved in the FTK project: apart from the development of a data format optimised for ternary-bit encoding of track candidates, I was responsible for the testing and integration of one electronics board (AM board) in the full FTK chain at a setup at CERN.

General information on the LHC and the studied proton-proton (pp) collisions will be presented in Sect. 5.1 before the ATLAS design and its sub-detectors are

---

[1] During the LHC design, it was in fact the strength of available magnets that determined the maximal energy, given the size of the already existing tunnel.

introduced in Sect. 5.2 along with its trigger system (Sect. 5.3). Subsequently, an overview of the relevant physics object definitions is given in Sect. 5.4 and event simulation is introduced in Sect. 5.5.

5.1 Particle Collisions at the LHC

The Large Hadron Collider (LHC) is currently the world's largest and most powerful particle accelerator. It is located at the European Organisation for Nuclear Research (CERN[2]), close to Geneva, Switzerland. The accelerator is designed to provide pp collisions at a centre-of-mass energy of up to 14 TeV. It can also produce Pb–Pb and p–Pb collisions. Up to now, the LHC delivered pp collisions at centre-of-mass energies of 7, 8 and 13 TeV. Small datasets at collision energies of 900 GeV, 2.76 and 5 TeV were also provided. Pb–Pb collisions at centre-of-mass energies of 2.76 and 5.02 TeV per nucleon and p–Pb collisions with 5.02 TeV centre-of-mass energy per nucleon were delivered as well. The LHC's first run, *Run I*, lasted from 2009 until March 2013. After a scheduled, long shutdown, the second run, *Run II* started in 2015 and will be ongoing until the end of 2018. The LHC is installed in an about 27 km long tunnel between 45 and 170 m underground, formerly accommodating the Large Electron Positron collider (LEP) where the electroweak W and Z bosons were discovered. The proton beams are bent by 1232 superconducting dipole magnets providing fields of up to 8 T, while multipole magnets focus the beams. Two separate beam pipes at ultra-high vacuum are contained in the magnets, in which the beams travel in opposite directions. In order to sustain the superconductivity of the magnets they need to be cooled down to 1.9 K, using liquid helium.

Four large experiments are placed at interaction points of the LHC where the beams can be brought to collision. ATLAS (A Toroidal LHC ApparatuS) and CMS (Compact Muon Solenoid) are multi-purpose detectors featuring extensive semi-conductor based tracking systems, large-coverage calorimeters and efficient muon detectors. They were optimised for the discovery and measurement of the Higgs boson and the search for new physics, but also pursue a considerable program of Standard Model measurements and heavy-ion physics. ALICE (A Large Ion Collider Experiment) focuses on the study of heavy-ion collisions, relying on a large and very performant Time Projection Chamber (TPC) in addition to other sub-detectors. Since the decay products of B-hadrons are often expected to be found in the forward region, LHCb (LHC beauty), an experiment designed for the precision study of flavour physics and CP violation, is built as a one-sided forward spectrometer. Both LHCb and ALICE require less collisions per time, achieved by defocussing or separating the beams before the collision.

The LHC is part of a whole accelerator complex and its experiments are only a subset of those operating at the CERN facilities, as detailed in Fig. 5.1. Protons

---

[2]The abbreviation CERN originates from the french name Conseil Européen pour la Recherche Nucléaire.

5.1 Particle Collisions at the LHC

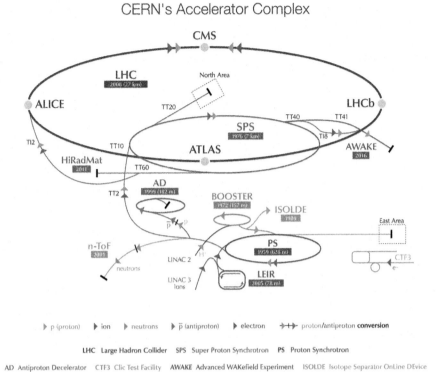

Fig. 5.1 Sketch of CERN's accelerator complex, including the LHC [2]

entering the LHC are produced by ionising hydrogen atoms. Then they are accelerated to 50 MeV with the Linac-2. The Proton Synchrotron Booster, the Proton Synchrotron (PS), and the Super Proton Synchrotron (SPS) accelerate them further to 450 GeV. The protons are eventually injected into the LHC in opposite directions and accelerated further. The protons form spatial bunches which were separated by 50 ns during the 8 TeV data taking and by 25 ns for the 13 TeV data taking. Trains of bunches are formed and a so-called abort gap between them allows for the beam dump mechanisms to act if needed.

The expected event rate of a certain process is determined by the product of its cross section and the so-called *instantaneous luminosity*: $R = \sigma \cdot L$. Hence, the luminosity is measured as $[L] = \text{cm}^{-2}\,\text{s}^{-1}$. It is defined as:

$$L = \frac{N_p^2 n_b f \gamma}{4\pi \epsilon \beta^*} F \rightarrow \frac{\text{crossing frequency} \cdot N_{\text{protons in beam 1}} \cdot N_{\text{protons in beam 2}}}{\text{beam overlap}}. \quad (5.1)$$

N_p denotes the number of protons per bunch, n_b the number of bunches per beam, f the revolution frequency and γ the relativistic factor. The numerator gives hence the number of interactions per time interval. In the denominator, the beam cross sectional size at injection, β^*, and the beam emittance ε are used to describe the area of overlap between the two colliding beams. The factor F can account for a possible beam crossing angle. The LHC was designed to reach peak luminosities of $L = 10^{34} \, \text{cm}^{-2} \, \text{s}^{-1}$. During the 8 TeV data-taking, the bunch spacing was kept at 50 ns (corresponding to a bunch crossing rate of 20 MHz) instead of the design value of 25 ns (corresponding to a bunch-crossing rate of 40 MHz). However, instantaneous luminosities of almost the design value were reached even under these conditions by increasing the number of protons per bunch. During data taking in 2016, the design luminosity of pp collisions was exceeded. The amount of collisions produced over a certain time period is quantified by the *integrated luminosity*, $\mathcal{L} = \int L dt$ and is measured in units of inverse cross section.

In order to increase the capability of probing lower cross sections the luminosity is increased as much as possible by having up to 10^{11} protons per bunch and closely spaced bunches. This leads to the occurrence of simultaneous pp collisions, so-called *pile-up*. Several interactions taking place in one bunch crossing, referred to as *in-time pile-up*, requires that the experiments are able to distinguish the different interaction vertices and their associated particles. Additionally, signals originating from adjacent bunch crossings might overlap in slow detector components and in the read-out electronics, which is called *out-of-time pile-up*. Pile-up is either measured as the number of primary vertices, N_{PV}, or as the average number of interactions per bunch crossing, $\langle \mu \rangle$. N_{PV} does not consider out-of-time pile-up and is built from reconstructed vertices, while the calculation of $\langle \mu \rangle$ is based on the measured luminosity.

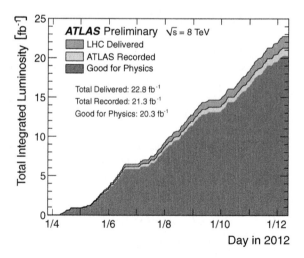

Fig. 5.2 Summary of the integrated luminosity recorded with ATLAS in 2012 [3]

5.1 Particle Collisions at the LHC

The *Monojet Analysis* that will be presented in Chap. 7 is based on 20.3 fb^{-1} of data from pp collisions at 8 TeV. The accumulated luminosity and the pile-up profile are shown in Figs. 5.2 and 5.3, respectively. The *Stop Analysis* that will be the subject of discussion in Chap. 9 uses the 13 TeV data that was collected in 2015 and 2016 until July 7. Figure 5.4 gives the luminosity summary for 2015 and 2016 and the pile-up distributions are shown in Fig. 5.5. In total, 13.2 fb^{-1} of 13 TeV pp collisions were considered for the work presented in this thesis.

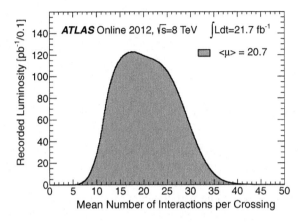

Fig. 5.3 Pile-up profile in 8 TeV pp collisions as measured with ATLAS in 2012 [3]

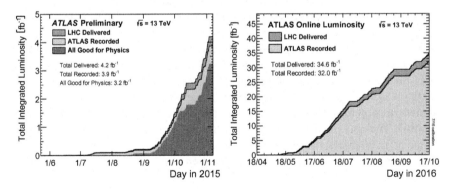

Fig. 5.4 Summary of the integrated luminosity recorded with ATLAS in 2015 [4] (left) and 2016 [4] (right)

Fig. 5.5 Pile-up profile in 13 TeV *pp* collisions as measured with ATLAS in 2015 and 2016 [4]

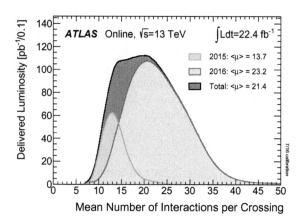

5.2 The ATLAS Detector

By studying in detail the collision products and their characteristics, information can be obtained on the underlying physics processes giving rise to the production of these particles. When particles traverse any material, they interact with it, resulting in a loss of energy of the traversing particle: either because the traversed material gets ionised or excited or because the particle emits radiation. Particle detectors function such that they convert this lost energy into an electronic signal that is recorded and can be analysed. Tracking detectors constrain the point of energy loss spatially in order to reconstruct the trajectory of the particle (*track*). If a magnetic field is applied, the bending of the track allows to constrain the particle's momentum and electric charge. Calorimeters are optimised to measure the energy of the traversing particle. By stopping it in the calorimeter material and measuring the released energy very precisely the original energy of the particle can be reconstructed.

ATLAS [5] is a general-purpose detector, designed to study many different aspects of modern particle physics. It is realised as a cylindrically symmetric magnetic spectrometer. The tracking detectors closest to the interaction point provide information on the particle trajectories and allow for an efficient vertex reconstruction, which is crucial for pile-up rejection and the identification of e.g. B-hadrons or tau leptons.

5.2.1 Coordinate System and Variable Definitions

The nominal interaction point where the proton beams are expected to collide defines the origin of the coordinate system used in ATLAS. The x-axis points towards the centre of the LHC, the z-axis is defined parallel to the beam circulating counter-clockwise and the y-axis to be orthogonal to both, such that a right-handed coordinate system is formed. The azimuthal angle ϕ in the x–y plane, defined relative to the

5.2 The ATLAS Detector

x-axis, and the polar angle θ in the x–z plane, relative to the z-axis, are used to denote coordinates. Since θ is not Lorentz-invariant, the pseudo-rapidity is often considered instead. It is defined as: $\eta = -\ln(\tan(\theta/2))$. The rapidity y is subsequently given by:

$$y = \frac{1}{2} \ln \frac{E + p_z}{E - p_z}, \qquad (5.2)$$

with E being the particle energy and p_z being its longitudinal momentum. In the limit of massless particles, the rapidity equals the pseudo-rapidity. The distance $\Delta R = \sqrt{\Delta \eta^2 + \Delta \phi^2}$ is often used to quantify how close two objects are to each other.

The transverse momentum p_T is defined as $p_T = \sqrt{p_x^2 + p_y^2}$, the transverse energy is given analogously. Since the incoming partons have no transverse momentum, the vectorial sum of transverse momenta of all produced objects in the collision has to be zero due to momentum conservation. This is not fulfilled experimentally if invisible or undetected particles are produced in the collision and the negative vectorial sum of the object p_T's is defined as missing transverse momentum $\vec{p}_T^{\,miss}$. Its amplitude is given by $E_T^{miss} = \sqrt{p_{x,miss}^2 + p_{y,miss}^2}$ and is called missing transverse energy.

The so-called transverse mass m_T targets leptonic W-boson decays. The expression $m_T = \sqrt{2 p_T^\ell E_T^{miss} (1 - \cos \Delta \phi(\ell, E_T^{miss}))}$ aims at reconstructing the mass of a common parent particle of neutrino and lepton. Since E_T^{miss} can only be defined in the transverse plane, the obtained transverse mass is a lower bound of the true parent mass.

5.2.2 Detector Design

The ATLAS detector is characterised by a powerful muon system, motivated by the aim to discover and measure the Higgs Boson[3] and calorimetry that covers almost the full solid angle of 4π, which allows to test multiple models of new physics, often characterised by large E_T^{miss}. ATLAS is built in layers, as it is typical for general-purpose particle detectors. The innermost layer is a tracking system surrounding the interaction point, immersed in a magnetic field. It is enclosed by electromagnetic and hadronic calorimeters as well as, in the outermost layer, a muon spectrometer. A schematic overview is given in Fig. 5.6. The detector can be sub-divided into the *barrel* and the *end-cap* region. The barrel is cylindrically symmetric around the beam pipe and typically extends up to $|\eta| < 1.4$. The endcaps "close" the open sides of the barrel by cylindrical structures, extending the range to up to $|\eta| < 5$. The following description of the detector and its sub-systems is largely based on Ref. [5].

---

[3] The Higgs decay channel $H \to 4\mu$ is very important, since it is extremely clean to reconstruct while being presented with very low Standard Model background.

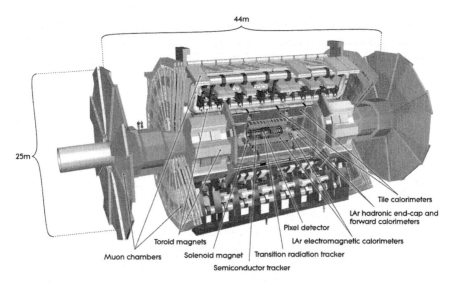

Fig. 5.6 Schematic view of the ATLAS detector and its sub-systems [6]

Inner Detector: The Inner Detector (ID) records information on the particle trajectories. It is surrounded by a superconducting solenoid magnet that provides a 2 T field in which the particle tracks are bent. It consists of four sub-systems. Closest to the interaction point, the Insertable B-Layer (IBL) was installed during the shutdown between Run I and Run II. It is a very-high-resolution semiconductor pixel detector, extending up to $|\eta| < 2.9$. In order to improve vertex reconstruction and B-hadron identification as much as possible, it was installed as close as possible to the interaction point, around a new, thin beam pipe at a radial distance of only 3.3 cm from the beam axis. This requires the sensors to be very robust against ionising radiation. The IBL is surrounded by the Pixel Detector, consisting of three layers of semiconducting pixels in the barrel region and three discs in each end-cap. It extends up to $|\eta| < 2.5$ between 5 and 12 cm radial distance from the interaction point. Its high granularity requires 80 million read-out channels and leads to a spatial resolution of $10\,\mu\text{m} \times 115\,\mu\text{m}$ in $R - \phi \times z$. A silicon microstrip detector, the Semi-Conductor Tracker (SCT), is located at radii between 30 and 51 cm from the interaction point in the region of $|\eta| < 2.5$. Each of the four barrel layers and 2×9 end-cap disks contains two sub-layers with tilted strip orientations, providing a spatial resolution of $17\,\mu\text{m} \times 580\,\mu\text{m}$ in $R - \phi \times z$. The outermost part of the ID is the Transition Radiation Tracker (TRT), a system of gas-filled straws that are parallel to the beam pipe in the barrel and radially oriented in the end-caps. It extends up to a radius of 108 cm and $|\eta| < 1.96$. By exploiting the difference in emitted radiation between electrons and other particle species when they traverse the material, it allows for a very good separation of electrons and other particle types, pions in particular. The lower spatial resolution ($130\,\mu\text{m}$) is compensated by the many hits provided per track (36) and the larger track length. An overview of the arrangement of the ID sub-

5.2 The ATLAS Detector

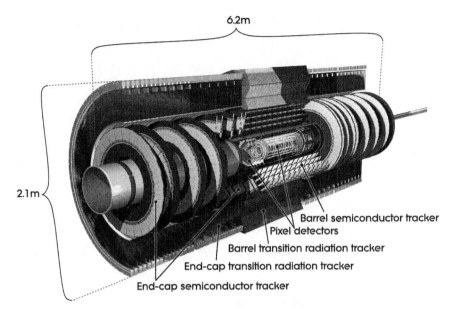

Fig. 5.7 Schematic view of the ATLAS inner detector [7]. The IBL (not shown here) is located between the beam pipe and the inner-most layer of the pixel detector

systems is given in Fig. 5.7. The ID ensures a precise tracking, enabling an efficient reconstruction of particle momenta and primary and secondary vertices. The latter is especially important to reject pile-up and in the identification of B-hadrons and tau leptons. A momentum resolution of $\sigma_{p_T}/p_T = (0.05\% \cdot p_T[\,\text{GeV}] + 1\%)$ [5] is targeted.

Calorimeter: An electromagnetic calorimeter, developed to contain and measure the showers of electrons and photons, is surrounded by a hadronic calorimeter. The electromagnetic calorimeter is built as a sampling calorimeter, meaning that alternating layers of absorbing and active material are used. Liquid Argon is used as active material and is combined with lead absorbers. The absorbers, as well as the electrodes are accordion-shaped to prevent detection gaps in transverse direction. Its thickness ranges between 22 (central) and 24 (forward) radiation lengths to ensure that the full electromagnetic shower is contained. In the barrel region it consists of two half-barrels with 16 modules each and extends up to radii of 4 m and $|\eta| < 1.475$. An additional *pre-sampler* layer is added right after the Inner Detector to estimate the energy that has been lost before the particles enter the calorimeter. The end-cap regions are equipped with eight wedge-shaped modules each. It is designed to achieve an energy resolution of $\sigma_E/E = (10\%/\sqrt{E[\,\text{GeV}]} + 0.7\%)$ [5]. The high granularity in the first layers of this calorimeter and its longitudinal separation allows for a reconstruction of the photon direction and to disentangle close-by photons.

The outer hadronic calorimeter combines scintillating tiles with steel absorbers. It reaches a thickness of ten interaction lengths and is hence able to fully stop

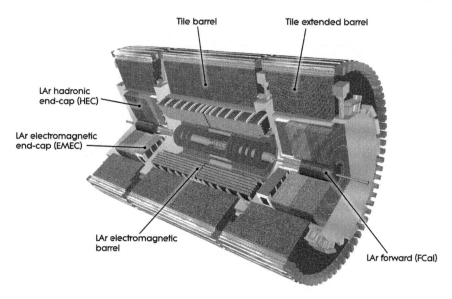

Fig. 5.8 Schematic view of the ATLAS calorimeters [8]

particles up to energies of several TeV.[4] The tile calorimeter covers $|\eta| < 1.7$ and is supplemented at larger pseudo-rapidities of up to $|\eta| < 3.2$ by the hadronic end-cap calorimeter (HEC), a copper-liquid-Argon calorimeter, consisting of two discs per end-cap. An energy resolution of $\sigma_E/E = (50\%/\sqrt{E[\,\text{GeV}]} + 3\%)$ [5] is aimed for.

The forward calorimeter covers $3 < |\eta| < 4.9$ and is composed of copper-tungsten as absorber and liquid Argon as active material, combining electromagnetic and hadronic calorimetry. The achieved energy resolution is expected to be: $\sigma/E = (100\%/\sqrt{E[\,\text{GeV}]} + 10\%)$ [5].

The calorimeter system with its sub-detectors is sketched in Fig. 5.8.

Muon Spectrometer: The function of the ATLAS muon system is twofold. It provides precise measurements of trajectories of muons as well as trigger signals for events containing muon candidates. The muon momentum measurement is based on the bending of tracks in the field of superconducting toroid magnets. This is ensured in the range of $|\eta| < 1.4$ by the barrel toroid, providing a field of up to 0.5 T, between $1.6 < |\eta| < 2.7$, the smaller end-cap toroids provide a magnetic field of up to 1 T. Each toroid consists of eight coils, arranged with an eight-fold azimuthal symmetry around the calorimeters. While the solenoid around the ID causes a bending in the transverse plane orthogonal to the beam pipe, the toroids deflect the muon tracks in the longitudinal direction. The three cylindrical layers of precision tracking chambers consist of so-called Monitored Drift Tubes (MDTs), supplemented by Cathode Strip Chambers (CSCs) in the forward region beyond $|\eta| > 2.7$. Due to support structures,

---

[4]For very energetic particles the energy might not be fully contained in the calorimeter in some cases and a signal in the muon system is observed, which is called *punch-through*.

5.2 The ATLAS Detector

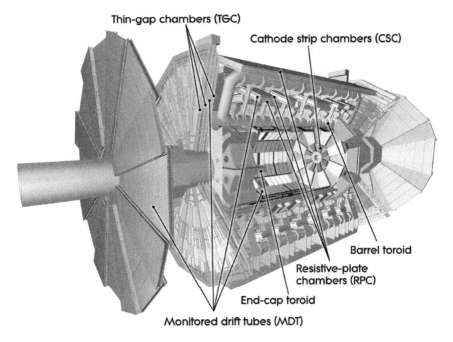

Fig. 5.9 Schematic view of the ATLAS muon system [5]

there is an uninstrumented gap at $\eta \sim 0$. The trigger chambers need to operate fast and rely on Resistive Plate Chambers (RPCs) and Thin Gap Chambers (TGCs). The muon system extends up to a radius of over 20 m from the interaction point and is the largest sub-detector by volume. An overview is given in Fig. 5.9. The design muon momentum resolution is $\sigma_{p_T}/p_T = 10\%$ at a muon p_T of 1 TeV [5].

5.3 Data Acquisition and Trigger

The proton bunches in the LHC cross every 50 (25) ns in Run I (Run II), at a rate of 20 (40) MHz. Only a small fraction of the collision events contain physics processes that are of interest for the analyses. The ATLAS detector cannot be read out sufficiently fast to record every event. Furthermore, bandwidth and data storage are limited. It is hence necessary to implement a triggering system that performs a basic decision of whether an event seems interesting enough to be recorded. It must be ensured that the trigger algorithms reliably cover all relevant physics scenarios and do not introduce a bias of any kind, while a balance has to be found between the coverage and the rate.

The rate reduction is performed in two steps. The first step, Level 1 (L1), needs to provide a decision very quickly, namely after 2.5 µs. To achieve this, the decision is purely hardware-based, using custom-built electronics. The Central Trigger Processor (CTP) combines information provided by the calorimeter and the muon chambers to identify interesting events. If the event is accepted at L1, Regions Of Interest (ROIs), built around relevant objects that were identified by the CTP, are communicated to the second level of the trigger, called High-Level Trigger (HLT), which operates software-based. For Run I, an intermediate step was defined at this stage, namely Level 2 (L2): here, based on the information in the ROIs, the rate was further reduced down to a few kHz. This level was omitted in Run II, any event passing the L1 is directly processed by the last step of the trigger, the Event Filter (EF). There, the event is analysed using the full event information. While this allows to lower the rate down to 300–500 Hz, these decisions are made on the order of seconds.

The trigger object reconstruction and the trigger selection requirements for every step are combined into so-called trigger chains. These trigger chains are collected in the trigger menu that defines which triggers are applied during run time. In some cases, triggers that would lead to very large rates but are needed e.g. for calibration and validation purposes are pre-scaled, i.e. only a randomly selected fraction of triggered events is recorded.

During Run I, based on which trigger accepted an event it was assigned to one or more data streams, e.g. the *muon stream*, the *electron-photon stream*, etc. This reduced the amount of data that needed to be processed by the analysers. In Run II, a similar reduction was achieved by the definition of analysis derivations: for each group of analyses some basic selection was applied and the resulting reduced data format then provided to the analysers. Events for which the trigger decision took unusually long are recorded in the so-called debug stream.

The performance of a trigger chain is typically studied in a dataset that is selected by another orthogonal or looser trigger. This allows to study the trigger efficiency, especially in the so-called turn-on region before the efficiency plateau is reached. This method of determining the trigger efficiencies is called *boot-strapping*. In many cases, also alternative approaches can be followed, for example the tag-and-probe strategy. This method relies on the selection of a certain event topology called tag, e.g. $Z/\gamma^*(\to e^+e^-)$ events which were triggered by an electron trigger. The event selection ensures that the second electron of the event, the *probe*, should have been triggered as well. Hence, the trigger efficiency can be estimated from how often this is actually the case. Generally, trigger efficiencies are determined as a function of the object p_T and η.

5.3.1 Missing Transverse Energy Trigger

Both analyses presented in the following rely on an E_T^{miss}-based trigger [9, 10]. The transverse momentum used in the trigger system is calculated from calorimeter-based global energy sums, possibly supplemented with information from the muon system.

The L1 trigger uses firmware on custom electronics to sum the energy deposited in coarse-grained projective *trigger towers* with $\Delta\eta \times \Delta\phi \sim 0.2 \times 0.2$ for $|\eta| <$

2.5 (coarser for larger η) to determine the total transverse momentum. The energy resolution and the level of zero suppression for trigger towers are found to be about 1 GeV. At the next trigger level (L2), the firmware-based sum of energy observed in groups of about 128 calorimeter cells is considered. Cells with energies less than three times the noise standard deviation σ_N as well as noisy cells are suppressed from the energy sum to reduce the effects from fluctuations. The last trigger step (EF) exploits the full calorimeter granularity. Two strategies are applied. Either, the energy deposited in all the cells is summed, where cells with $|E| < 2\sigma_N$ are omitted in the algorithm and the trigger was protected from large fluctuations by rejecting cells with $E_i < -5\sigma_N$. Or the cluster algorithm uses seed cells with $|E_i| > 4\sigma_N$, surrounding cells with $|E_i| > 2\sigma_N$, and all immediate neighbour cells to calculate the E_T^{miss} and applied a local hadronic calibration, though not the full object calibration of the reconstruction algorithms.

The trigger logic was very similar for the Run I and Run II datasets which are analysed in the following, apart from the omission of L2 in Run II [11]. Many improvements of the E_T^{miss} triggers are underway, especially in view of pile-up subtraction techniques [12]. They were not yet applied for the analysed dataset of 13 TeV collisions.

5.3.2 The Fast TracKer (FTK)

Tracking at trigger level is beneficial to limit the trigger rate in high-pileup conditions while maintaining a good efficiency for relevant physics processes. However, tracking generally takes long and is only included in the trigger decision within restricted regions of the event or for tracks seeded by other identified objects. The Fast TracKer (FTK) [13] provides a solution: for every event passing the L1 trigger, it performs a very fast hardware-based tracking for the whole event and transfers the track information to the HLT where this can be included in the algorithms.

Including the information provided by FTK in the HLT decisions leads to higher trigger efficiencies for medium-p_T b-quarks and tau leptons with high background rejection, since both b-tagging and tau lepton identification rely on track information: b-jets are characterised by a displaced vertex that can be reconstructed from the tracks in the event and jets coming from hadronic tau decays have significantly less tracks in a smaller cone than standard jets [13]. This is especially important for Higgs coupling measurements, where third generation leptons play a crucial role, as well as for SUSY searches, since scenarios with light stops, sbottoms or staus are interesting but challenging to rule out or discover [13]. Furthermore, the primary vertex and the pile-up condition of an event can be determined using FTK tracks. Many trigger algorithms can be improved when including this information in the HLT decision, especially when relying on isolation variables (such as lepton triggers) and calorimeter information (such as jet and E_T^{miss} triggers), since they are both affected by pile-up effects.

The hardware-based tracking performed by FTK is based on full-precision hit information from all channels of the ATLAS silicon detectors. The resulting tracks are sent to the HLT to be used in the software algorithms. In order to cope with event rates of up to 100 kHz the tracking performed by FTK has to be several orders of magnitude faster than tracking at reconstruction level. Hence, the processing of the data is organised as parallel as possible: the signals from the full detector volume are split into 64 regions, so-called *towers*, which are processed independently. Further, the data volume is decreased as much as possible by a custom clustering algorithm defining *hits* which are considered later on instead of the full pixel/strip information. In addition, the hit information is re-binned into coarse-resolution *superstrips* whenever appropriate by grouping several pixels or strips together.

FTK performs the tracking in two steps. At first, track candidates are identified by comparing the fired superstrips to predefined trajectories stored in memory. Such predefined *patterns* refer to a list of superstrips crossed by the trajectory of a simulated particle as it traversed the detector layers. The found track candidates at coarse resolution (roads) seed a full-resolution track fitting performed in Field-Programmable Gate Arrays (FPGAs). Only considering hits within these roads reduces the combinatorics significantly and hence makes the fit itself much faster. The pattern matching procedure is based on the use of custom associative memory (AM) chips designed to perform pattern matching at very high speed (about 100 MHz). It allows to compare the incoming data simultaneously to all stored patterns.

The data flow and the components of FTK are shown schematically in Fig. 5.10. Starting from the Input Mezzanine cards which receive data from the tracking detectors and perform the custom pixel/strip clustering algorithm on FPGAs, the hits are sent to the 32 Data Formatters, responsible for the geometrical grouping of the data into the 64 independent $\eta - \phi$ towers. For each tower, the information is distributed to the corresponding processing unit, which consists of two Auxiliary (AUX) cards and their Associative Memory Boards (AMB). The full-resolution hits are reclustered into coarse superstrips by the AUX Data Organisers which are then communicated to the corresponding AMBs, where the AM chips match the incoming superstrips to the stored track patterns. The found track candidates are input to the AUX Track Fitter which performs a first tracking within the roads, relying on the full-resolution hit information in eight out of the 12 silicon detector layers. Subsequently, the AUX Hit Warrior function removes duplicated tracks. Based on the results, the 32 Second Stage Boards extend the fit to all layers of the Tracker and refine the track parameters in a second fit. Finally, two interface cards (FLICs) take care of the communication with the HLT. In summary, eight full nine-unit VME crates and five ATCA shelves host about 2000 FPGAs and 8000 custom AM chips, which makes the FTK a very complex custom parallel supercomputer.

The parameters of the pattern matching have to be optimised: while narrow roads permit a fast track fitting, efficient matching would require to store many patterns in the AM. Wide roads, on the other hand, allow for fewer patterns stored in memory but the increased combinatorics within the matched roads slow down the track fitting. This choice is optimised by implementing the feature of variable resolution of the

5.3 Data Acquisition and Trigger

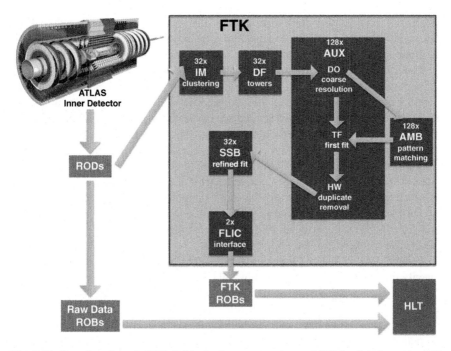

Fig. 5.10 Data flow through FTK. IM is the input mezzanine, and DF the data formatter. DO is the data organiser, TF is the track fitter, and HW is the hit warrior, which are all parts of the AUX (auxiliary board). AMB denotes the associative memory board. The second stage board is abbreviated by SSB, FLIC stands for FTK-to-Level-2 interface crate. ROB is the ATLAS read out input buffer, ROD is a silicon detector read out driver

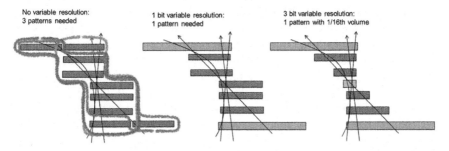

Fig. 5.11 The sketches indicate how variable resolution can reduce the number of patterns stored in memory and the contribution from random hits [13]. The coloured lines show the patterns needed to accept the tracks. The eighth ID layer (IBL) is not included here

roads via ternary bits in the AM logic [14], as illustrated in Fig. 5.11. Furthermore, the number of required matching layers is programmable.

The pattern banks are generated once from single track MC samples in the finest resolution possible (superstrips), where the hit position is indicated by the superstrip

ID (SSID). Each pattern consists of eight SSIDs, corresponding to the eight silicon layers. In following iterations, the resolution can be optimised: the feature of variable resolution is implemented via ternary bits which are called "Don't Care" (DC) bits.

The SSID format consists of a part denoting the module ID and a part indicating the superstrip position in x and y. The ternary bits of the x and y positions are grouped together into the least significant bits of the SSID. Gray encoding [15] is applied to the least significant bits. In Gray code, two consecutive numbers only differ by one digit. If one of the possible ternary bits is declared as a DC bit, the Gray code ensures that, in practice, two consecutive superstrips get combined into one larger one. For example, superstrip 110 and 111 are supposed to be next to each other. If the least significant bit is declared to become a ternary bit, the SSID 11X denotes a superstrip of twice the original size, comprising both superstrips 110 and 111. In hardware, two bits are reserved for one ternary bit with the assignment of "0 = 01", "1 = 10" and "X = 00".[5] I was partly responsible for implementing this SSID format into the simulation software of FTK.

In the following, the test procedure for the AM boards is described, in which I was heavily involved. The data flow within and between the different electronics boards of FTK as well as the agreement of the actual hardware system and its simulation is tested by using so-called *test vectors*. Such test vectors consist of pseudo input data in a format that can be interpreted by both hardware and simulation. In this way, the outputs from hardware and simulation can be compared and should be identical. To this end, small pattern banks were prepared, adapted for the few considered input test events to compare fired superstrips, found roads and matched pattern IDs in hardware and simulation. For these dedicated pattern banks I developed a special format that is independent of the (huge) software package for FTK and can hence be easily loaded on the boards. This format was the starting point for a redefinition of the pattern bank format used FTK-wide.

A single AM board, equipped with only one instead of four mezzanine boards has been tested using the above-described test vectors in the infrastructure provided at CERN. After perfect, bitwise agreement was achieved between the output of the electronics board and the simulation, the data chain was extended to precedent and following boards. Most importantly the communication with the AUX had to be correctly established. After this was ensured, the command-line configuration and execution procedure was implemented in the ATLAS software used to steer detectors and components at run time, "Run Control" [16]. Problems encountered and solved during the tests were for example due to an inconsistent threshold (voltage) for the signal stopping the sending of data to the subsequent board, leading to lost data. Furthermore, duplicated events occurred due to an inconsistent number of idle words between the different AMB algorithms. The successful establishment of the communication between AMB and AUX within the Run Control environment was the basis for AMB integration in the FTK system installed at the electronics cavern of ATLAS.

---

[5]"11" is not defined.

5.3 Data Acquisition and Trigger

Despite the huge increase in speed, the quality of FTK tracks is in many respects comparable to fully reconstructed tracks. The momentum and angular resolution is only slightly worse and small effects from pile-up enter. As an example, a comparison of the transverse impact parameter is shown in Fig. 5.12. Also the b-tagging efficiency is found to be similar to reconstruction level in simulation, while maintaining a high light-jet rejection. The number of vertices found in the reconstruction and in FTK correspond linearly. More details can be found in the FTK Technical Design Report [13]. With FTK, it is possible to efficiently trigger on one-prong tau decays as seen in a $H \rightarrow \tau\tau$ simulation. In particular low-p_T tau candidates can be recovered when including FTK tracks in the trigger algorithms. This is shown in Fig. 5.13.

While Run II of the LHC is ongoing, the first boards of FTK are being installed, already reading ATLAS data in parasitic mode. The full processing of the complete barrel region is expected for the end of 2016, after which the complete integration within HLT and the extension to full coverage will follow. In view of the HL-LHC,

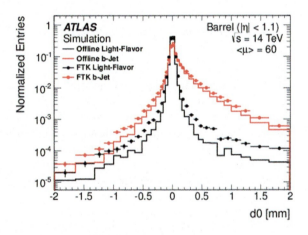

Fig. 5.12 Comparison of the transverse impact parameter d_0 between FTK (points) and reconstruction-level tracks (histograms) in the barrel region [17]

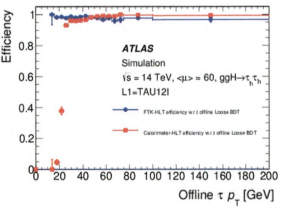

Fig. 5.13 Identification efficiency for tau leptons as a function of reconstructed tau p_T applying an FTK-based selection (blue) and a calorimeter-based selection (red) at HLT [18]. The efficiency is defined as the fraction of Level-1 tau matched to a reconstructed tau [19]

the FTK concept is discussed to be extended to cope with luminosities higher than the design goal of the current system. Also ideas of using tracking information already at Level 1, involving an upgraded AM chip, are considered [20].

5.4 Physics Object Definitions

Interactions of the particles produced in the collision with the detector material leave traces that are recorded in form of electronic signals. In order to be able to analyse the collisions these signals have to be processed, grouped and combined such that they can be interpreted as physics objects. This step is called *reconstruction*. The most important physics objects for the later presented analyses and how they are built from detector signals will be detailed in the following.

5.4.1 Track and Vertex Reconstruction

The reconstruction of particle tracks requires a minimum number of hits in the silicon detectors of the ID as seed[6] and then extends the track by adding additional hits, following either a Gaussian sum filter or a global χ^2 fit procedure. "Fake" tracks from instrumental effects or pile-up are reduced by requiring a hit in the IBL or the innermost pixel layer and by rejecting tracks where an expected hit in an intermediate layer is not found. In order to be considered as a track, $p_T > 400$ MeV and $|\eta| < 2.5$ has to be satisfied. There exist also several other types of tracks that are reconstructed differently, e.g. only from TRT information or only from hits in the muon chambers and are used in special cases.

Possible vertex candidates are identified by extrapolating the found tracks to the beam line. For each candidate within the three-dimensional beam spot position, all tracks within 7σ of the vertex candidate are taken to originate from it and fitted via an iterative χ^2 procedure. Based on the resulting χ^2 a weight is assigned to each track.

Since physics processes of interest often have a large number of high-p_T tracks, the vertex presenting the largest sum of the squared track p_T's is taken to be the primary one. In further steps, secondary vertices and photon conversions are reconstructed by additional, dedicated algorithms.

The performance is found to be similar between Run I and Run II. The track reconstruction efficiency above $p_T > 5$ GeV reaches 85% or 90%, depending on the desired fake rejection working point [21]. The vertex reconstruction efficiency is around 90% for vertices having at least two associated tracks and decreases with increasing pile-up [22].

---

[6]The exact number depends on the desired track quality.

5.4.2 Muons

For the reconstruction of muon candidates, information from both the ID and the muon system (MS) is considered and is complemented with calorimetric information around $\eta \sim 0$, where the MS has a small gap [23, 24]. Track segments are built in each layer of the MS and combined into tracks through the full MS. ID tracks fulfilling some quality requirements are then combined with the MS tracks and form different muon types: *standalone muons* are fully based on the MS track that is extrapolated to the primary vertex, *combined muons* use both ID and MS tracks, *segment-tagged muons* rely on an ID track that is extrapolated to the MS and can be successfully matched to an energy deposit in at least one MS segment. Lastly, *calorimeter-tagged muons* have an ID track matched to a calorimeter cluster consistent with a minimally-ionising particle. While the combined muons are the cleanest ones, the fallback to other types increases the acceptance. The reconstruction efficiency is close to 99% in the central region.

Different identification working points are defined from *loose* to *tight* with decreasing signal efficiency and increasing background rejection, as described in Refs. [23, 24]. The criteria consider the quality of MS and ID track, including hit requirements, and how well they match. In Run I, the reconstruction efficiency is above 99% and the momentum resolution reaches 1.7% for central muons with $p_\mathrm{T} > 10$ GeV [23]. In Run II both the reconstruction efficiency and the momentum resolution are found to similar [24].

5.4.3 Electrons and Photons

Electrons, as well as photons, deposit all their energy in the electromagnetic (EM) calorimeter, such that no signal is expected in the hadronic calorimeter or the MS. In contrast to photons, electrons, as charged particles, leave tracks in the ID and can hence be distinguished. However, due to bremsstrahlung of electrons and conversions of photons into electron-positron pairs, their signatures are not easily separable.

The reconstruction relies on energy deposits in the cells of the electromagnetic calorimeter. *Towers* of $\Delta\eta \times \Delta\phi = 0.025 \times 0.025$ are used, which total transverse energy is determined as the sum of the energy deposited in all cells of the tower in all longitudinal layers. An algorithm tries to group several calorimeter cells in which energy is deposited together into so-called *clusters*, using towers with a total transverse energy above 2.5 GeV as seeds.

If a high-quality track originating from the primary vertex, reconstructed in the ID and extrapolated to the EM calorimeter, can be matched to the calorimeter cluster, the object is considered to be an electron. If no track is found, it is interpreted as an unconverted photon. If a matching track is found that is consistent with originating from a conversion vertex, a converted photon is assumed.

The object energy is estimated from rebuilt clusters or from track properties, depending on the pseudo-rapidity of the object, considering simulation-based correction factors. Additional calibration factors are derived in a tag-and-probe procedure from $Z/\gamma^*(\to e^+e^-)$ events in data. More information about this chain can be found in Ref. [25].

Electrons face backgrounds from wrongly-identified charged particles such as pions or jets that fake an electron signature. The further background is reduced by including transition-radiation information from the TRT. Different identification working points are defined from *loose* to *tight* with decreasing signal efficiency and increasing background rejection. They consider extension and shape of the EM shower, the fraction of energy deposited additionally in the hadronic calorimeter, the quality of the matched track and the matching requirements of track and cluster. Also the object isolation is considered in the identification. Details can be found in Ref. [26] for Run I and in Ref. [27] for Run II. Observed electron identification efficiencies range between 77% for tight identification criteria and 93% for loose identification criteria in an E_T range between 20 and 50 GeV.

For photons, a similar strategy is employed. The details of the identification criteria can be found in Ref. [28]. The photon identification efficiency increases from 53–64% at 10 GeV to 88–92% at 100 GeV [29].

5.4.4 Jets

Objects carrying a QCD colour charge only occur in colour-neutral bound states, resulting from an approximately linear increase of the QCD coupling strength when increasing the distance of interacting objects. This phenomenon is referred to as *confinement*. Therefore, the quarks and gluons that are produced in the hard parton scattering undergo a hadronisation process[7]: each of the quarks/gluons causes the formation of a spray of colour-neutral hadrons in its direction, where the abundance and momenta of the produced hadrons depend on the energy of the quark/gluon. Such a collimated spray is referred to as *jet*. Strictly speaking, the object *jet* is defined by– and depends on–the jet reconstruction algorithm and its parameters. The aim is to reconstruct the momentum, energy and direction of the original parton as closely as possible. In the analyses presented in this thesis jet candidates are reconstructed using the so-called *anti-k_t* algorithm [30]. The distance measure used in this algorithm reads:

$$d_{ij} = \min(k_{t,i}^{2p}, k_{t,j}^{2p}) \frac{\Delta_{ij}^2}{R^2}, \qquad (5.3)$$

where $\Delta_{ij} \equiv (y_i - y_j)^2 + (\phi_i - \phi_j)^2$, k_t is the transverse momentum, y denotes the rapidity of the constituents and R the radius parameter. The parameter p takes the

---

[7]The–compared to other quarks–very heavy top quarks are an exception: they decay before the hadronisation takes place.

5.4 Physics Object Definitions

value -1 for the anti-k_t algorithm, where its negative sign denotes that the jet reconstruction proceeds from the hardest to the softest objects. The similar k_T algorithm uses $p = 1$ and hence leads to a reconstruction from the softest to the hardest objects. The above-defined distance d_{ij} is compared to:

$$d_{iB} = k_{t,i}^{2p}. \tag{5.4}$$

If $d_{iB} > d_{ij}$, then the algorithm stops, the object j is not grouped with the object i and object i is considered to be a jet and removed from the list of inputs to further jets.

It is important to ensure that the jet algorithm leads to *infrared* and *collinear* safe definitions, i.e. that the final jets are insensitive to the addition of infinitely soft partons or to the splitting of one parton into two. The radius parameter has to be chosen such that on one hand the whole spray of particles is contained in the jet and that on the other hand the jets are not strongly affected by the underlying-event activity. The value of $R = 0.4$ is used for the analyses presented in this work.

Topologically connected calorimeter clusters (topo-clusters) are inputs to the jet algorithm. They are built from cells in which an energy deposit over a certain noise threshold is detected and grouped with neighbouring cells measuring a sufficiently high signal. The topo-cluster energy is found by summing the contributions from the individual cells. It is given at the electromagnetic (EM) energy scale, denoting the scale for which electromagnetic showers are correctly measured. In order to account for the difference in calorimeter response to electromagnetic or hadronic showers, the topo-clusters undergo a *local cluster weighting* (LCW), where the correction factor depends on whether they seem to be mostly hadronic or electromagnetic.

After the jets are reconstructed from the topo-clusters, further calibration and corrections are applied to the jet energy scale (JES) [31, 32]. At first, the energy estimated to come from pile-up contributions is subtracted. Then, the jet is constrained to originate from the reconstructed primary vertex instead of the nominal interaction point. The jet energy response determined as:

$$\mathcal{R} = \langle \frac{E_{reco}^{MC}}{E_{truth}^{MC}} \rangle \tag{5.5}$$

is inverted and applied as an energy- and η-dependent correction factor. At last, jets in data are corrected following in-situ energy measurements of well-balanced objects.

Jets are reconstructed above a transverse momentum of 7 GeV.

5.4.5 B-Jets

Due to the non-zero decay length of B-hadrons, jets originating from a b-quark can be discriminated from those from gluons or other quarks. Specific properties of the jet such as its impact parameter or associated secondary vertices are used in a multivariate approach to determine a discriminant between zero and one. In ATLAS, the so-called MV2c10 algorithm [33, 34] is applied for Run-II data.[8] Depending on the b-jet identification efficiency and the light-jet rejection desired for the analysis, an appropriate cut on the discriminant has to be defined. The b-jets entering the *Stop Analysis* described in Chap. 9 are tagged, if the b-jet discriminant exceeds a threshold that was chosen to obtain an identification efficiency of 77% of b-jets above 20 GeV in $t\bar{t}$ events. The corresponding light-jet rejection is 134 [34].

The b-tagging efficiency is found to vary only slightly with η but has a dependency on the jet p_T, where it is highest for jet p_T's around 100 GeV. In particular, it decreases for higher jet p_T's. From the b-tagging efficiency observed in data, correction factors are determined which are applied to simulated samples.

5.4.6 Hadronically Decaying Tau Leptons

Candidates for hadronically decaying tau leptons are seeded from calorimeter clusters that are reconstructed as a jet. The following variables are used to discriminate hadronic tau decays against jets. The central energy fraction gives the ratio of transverse energy in $\Delta R < 0.1$ and in $\Delta R < 0.2$ of the tau candidate. This is especially high for 1-prong decays and still higher than for quark or gluon jets in 3-prong decays. Similarly, the fraction of the jet momentum that is carried by the leading track is discriminating. Furthermore, the p_T-weighted distance of associated tracks to the tau candidate direction is considered; this is expected to be smaller for hadronic tau jets. Also the number of tracks in a distance of $0.2 < \Delta R < 0.4$ from the tau candidate is expected to be small for tau jets. In addition, the significance of the impact parameter of the leading track with respect to the estimated tau vertex, the maximum distance of any track associated with the tau candidate and its direction and the significance of the decay length with respect to the estimated tau vertex and the track mass are included. A dedicated π^0 identification allows to also construct and exploit the mass of the π^0 and the tracks, the number of π^0's and the ratio of the track and π^0 momentum and the calorimeter-based momentum as discriminating variables.

All these parameters feature in the training of a boosted decision tree (BDT). For each the one-prong and the three-prong decay, a separate BDT is trained. Its discriminant is used to define working points with different purity and efficiency, as described in Refs. [35, 36]. The efficiencies are by construction independent of pile-up and of tau p_T. They range from 60 to 45% for one-prong and from 50 to 30% for three-prong decays [31].

---

[8]Since b-jets within this thesis are only used in the *Stop Analysis* of Run-II data, the details of b-tagging in Run I are not discussed here.

5.4.7 Missing Transverse Energy

Many signatures of new physics involve additional undetectable particles and hence rely on E_T^{miss} as a discriminating variable. The E_T^{miss} is calculated as the negative vectorial sum of the transverse energy of all significant energy deposits in the detector and presents the momentum vector that would be required to balance the event. There are several prescriptions for the E_T^{miss} determination [37, 38]. The default formulation uses identified objects of electrons, photons, taus, jets and muons as input. The objects are required to pass a simple selection to reduce the fake contribution without being so tight as to throw away significant numbers of good objects, as both could bias the E_T^{miss} calculation. Each class of objects is calibrated and corrected separately. To avoid a double-counting of energy the potential overlap between objects needs to be resolved. In particular, electrons, photons, and taus are also reconstructed as jets and hence jets overlapping with such an object need to be excluded from the calculation.

Furthermore, energy deposits in the calorimeters and tracks which are not associated to one of these objects are considered in the so-called *soft term*. Here, the lower momentum threshold of tracks compared to calorimeter clusters and their better momentum resolution at low momenta is exploited. Tracks that could not be matched to any topo-cluster or reconstructed object are added to the calculation to account for low-p_T objects that did not reach the calorimeter or that are missed. If a track is matched to a topo-cluster, the track is considered in the calculation and the topo-cluster is discarded in order to profit from the better momentum resolution of tracks at low momentum. If a track is matched to multiple topo-clusters, the track enters the calculation as well as all topo-clusters but the on with the highest energy. All remaining topo-clusters are included in the soft term.

It is clear that E_T^{miss} is vulnerable to potential mis-measurements of energies and momenta–especially those of jets. The E_T^{miss} resolution in is etsimated to be between 5 and 30 GeV in simulations, depending on the total energy in the event.

5.5 Event Simulation

5.5.1 Sample Generation

Monte Carlo (MC) simulated samples [39] are used to develop selections that discriminate well between signal and background by studying their different kinematic behaviours. Simulated samples are also used to study detector acceptance and reconstruction efficiencies for signal and background processes and are needed in the background estimation. Furthermore, simulations allows to study systematic uncertainties from different sources in detail.

Given the large number of particles produced in hadron collisions, with momenta ranging over several orders of magnitude, the simulation of such events is not trivial. Especially the description of the non-perturbative soft QCD processes require a phenomenological approximation. The actual generation proceeds in several steps. First, the matrix element calculation of the hard scattering process between two incoming partons is calculated from perturbation theory (to some limited order), based on the relevant Feynman diagrams. If the simulation is performed at leading order (LO) in perturbation theory, the effect of next-to-leading-order (NLO) corrections can be parametrised by the k-factor, which is the ratio of the NLO to the LO calculation, often determined in a specific kinematic regime. A relatively small number of outgoing particles is produced in this process. Input to this step is the so-called parton distribution function (PDF), describing the constituents of the protons (sea and valence quarks, as well as gluons) and their momentum fraction at a given energy. Furthermore, the factorisation scale, marking the transition between the perturbative and non-perturbative regime, and the renormalisation scale for the running coupling of the strong interaction have to be fixed. Both scales are unphysical and their impact on the result decreases with each additional order of perturbation theory that is taken into account.

The evolution from the scale of the hard process down to the scale at which confinement takes place is described by parton shower algorithms that are based on the DGLAP [40–42] evolution equations. During this step, the incoming and outgoing partons *shower*, i.e. radiate off other partons or might split several times. The parton shower description accounts for higher order effects that are not included in the fixed order matrix element calculation of the hard process. The actual hadronisation step in which the partons get combined into colourless bound states is purely based on phenomenological models, such as the *colour string model* [43]. However, the parameters of the description do not depend on the actual hard process, meaning that they can be constrained based on a specific data set and then applied in other cases. It is important to also consider the *underlying event activity* arising from multiple parton interactions in the process. Pile-up events are usually included by overlaying the simulated event with several general pp collision events.

One difficulty in this chain is to consistently combine matrix element calculation and parton shower. Several procedures are applied, for example the so-called CKKW prescription [44]. Generally, some specific scale which marks the transition between the regimes needs to be fixed.

In the following, the most relevant generators used in this thesis are briefly discussed.

- MADGRAPH [45] is a leading-order matrix element generator that was extended to allow for the inclusion of loop diagrams [46]. MadGraph automatically generates Feynman diagrams and calculates matrix elements for user-specified processes. For the simulation of parton shower and hadronisation the output has to be transmitted to an external programme.
- SHERPA [47] is a general-purpose generator, covering both the matrix element calculation and the parton shower description. It is considered to be one of the most advanced programme for the automated generation of tree-level matrix elements for both Standard Model and new physics processes. It uses two matrix element generators that apply advanced phase-space integration techniques. Apart from

parton shower and hadronisation, SHERPA provides modelling of hadron and tau decays as well as electromagnetic final state radiation and the simulation of multi-parton scattering.
- PYTHIA [48–50] is also a general-purpose generator. It does not allow automated code generation for new processes, instead it provides more than 200 hard-coded subprocesses and is designed such that it facilitates the use of external input (e.g. from MADGRAPH for the simulation of parton showering, hadronisation and underlying event. PYTHIA includes elastic, single and double diffractive and non-diffractive soft processes, providing an inclusive description of the total pp interaction cross section. Hence, PYTHIA is also well-suited for the generation of pile-up events.
- HERWIG++ [51] generates hard processes, providing decays with fully considered spin correlations also for many models of new physics. Furthermore, it produces angular ordered parton showers and includes the modelling of hadronisation. The underlying event is inferred from multiple parton interactions and provided by a routine library called JIMMY [52]. HERWIG++ features sophisticated models for the decay of hadrons and tau leptons.
- MC@NLO [53] calculates hard processes at next-to-leading-order (NLO) in QCD. It provides an algorithm for parton showering and includes spin correlations for most processes. For the modelling of the underlying event, MC@NLO is typically interfaced to HERWIG++. MC@NLO is a specialised generator with a specific set of implemented processes, including Higgs boson, single vector boson, vector boson pair, heavy quark pair, single top, lepton pair and associated Higgs+W/Z production.
- AcerMC [54] is also a specialised generator, based on MADGRAPH, that is typically interfaced to PYTHIA or HERWIG. It is specifically designed to model Standard Model backgrounds, a library for the corresponding matrix elements and phase space modules is provided for several processes.
- ALPGEN [55] is a tree-level ME calculator, dedicated to the study of multi-parton hard processes in hadronic collisions with emphasis on configurations with high jet multiplicities. It is based on the evaluation of the relevant leading-order Feynman diagrams for strong and electroweak interactions. Calculations are provided for a list of specific final states.
- POWHEG [56] applies an advanced ME reweighting procedure, where the hardest interaction term is replaced by its NLO-weighted correspondent. It was extended to allow for the automatic implementation of any given NLO calculation. The algorithm does not depend on a particular parton shower and its output can be easily interfaced to any modern shower generator.

5.5.2 Detector Simulation

In order to connect simulated event data to detector signals from actual recorded data, the interaction of the particles with the ATLAS detector material and support structures needs to be simulated [57]. A detailed geometrical model of the detector is used within GEANT4 [58] to derive the energy deposits for each simulated particle. Subsequently, digitisation software specific to each sub-system converts the simulated energy deposits into electronics signals which would be observed in the detector [59]. After this step, the signals are reconstructed as for example tracks or calorimeter clusters analogous to collision data.

The detector simulation is time-consuming and takes up to several minutes for a typical event. Most of the time is spent by the GEANT4 simulation of the calorimeter response. For this reason, a faster simulation procedure was developed that applies a parametrised calorimeter cell response [60] instead of the above described *full simulation*. The average GEANT4 response to a given type of particle in a kinematic bin within the various calorimeters gets parametrised and look-up tables for energy deposits and interaction probabilities are provided. An improvement in processing time of up to an order of magnitude is achieved and this *fast simulation* (AFII) is found to be well suited if the highest level of precision in the calorimeter response is not required.

References

1. ATLAS Collaboration, Public collaboration website (2013), http://atlas.cern/discover/collaboration
2. F. Marcastel, CERN's Accelerator Complex. La chaîne des accélérateurs du CERN, https://cds.cern.ch/record/1621583
3. ATLAS Collaboration, Luminosity Public Results for Run I (2013), https://twiki.cern.ch/twiki/bin/view/AtlasPublic/LuminosityPublicResults
4. ATLAS Collaboration, Luminosity Public Results for Run II (2016), https://twiki.cern.ch/twiki/bin/view/AtlasPublic/LuminosityPublicResultsRun2
5. ATLAS Collaboration, G. Aad et al., The ATLAS experiment at the CERN large hadron collider. JINST **3**, S08003 (2008). https://doi.org/10.1088/1748-0221/3/08/S08003
6. ATLAS Collaboration, J. Pequenao, Computer generated image of the whole ATLAS detector, Mar 2008
7. ATLAS Collaboration, J. Pequenao, Computer generated image of the ATLAS inner detector, Mar 2008
8. ATLAS Collaboration, J. Pequenao, Computer generated image of the ATLAS calorimeter, Mar 2008
9. ATLAS Collaboration, The ATLAS transverse-momentum trigger performance at the LHC in 2011. Technical Report, ATLAS-CONF-2014-002, CERN, Geneva, Feb 2014, https://cds.cern.ch/record/1647616
10. A. Mincer, F.U. Bernlochner, A. Struebig, J. Schouwenberg, J.B. Beacham, The ATLAS transverse momentum trigger at the LHC. Technical Report, ATL-DAQ-PROC-2015-028, CERN, Geneva, Sep 2015, https://cds.cern.ch/record/2053969
11. ATLAS Collaboration, Performance of the ATLAS trigger system in 2015. Technical Report, ATL-COM-DAQ-2016-034, CERN, Geneva, Mar 2016, https://cds.cern.ch/record/2140103
12. A. Struebig, The ATLAS Transverse Momentum Trigger Evolution at the LHC Towards Run II, http://cds.cern.ch/record/2047799
13. M. Shochet, L. Tompkins, V. Cavaliere, P. Giannetti, A. Annovi, G. Volpi, Fast tracKer (FTK) technical design report. Technical Report, CERN-LHCC-2013-007. ATLAS-TDR-021, Jun 2013, http://cds.cern.ch/record/1552953

References

14. A. Annovi, S. Amerio, M. Beretta, E. Bossini, F. Crescioli, M. Dell'Orso, P. Giannetti, J. Hoff, V. Liberali, T. Liu, D. Magalotti, M. Piendibene, A. Sacco, A. Schoening, H.K. Soltveit, A. Stabile, R. Tripiccione, R. Vitillo, G. Volpi, A new "variable resolution associative memory" for high energy physics. Technical Report, ATL-UPGRADE-PROC-2011-004, CERN, Geneva, May 2011, http://cds.cern.ch/record/1352152
15. F. Gray, Pulse code communication, 17 Mar 1953, https://www.google.com/patents/US2632058. US Patent 2,632,058
16. ATLAS Collaboration, P. Jenni, M. Nessi, M. Nordberg, K. Smith, ATLAS high-level trigger, data-acquisition and controls: technical design report. Technical Design Report ATLAS. CERN, Geneva (2003), https://cds.cern.ch/record/616089
17. ATLAS Collaboration, FTK track transverse impact parameter (2016), https://twiki.cern.ch/twiki/pub/AtlasPublic/FTKPublicResults/can_OffAll_b_60_d0_signed_b.png
18. ATLAS Collaboration, FTK tau identification efficiency (2016), https://twiki.cern.ch/twiki/pub/AtlasPublic/FTKPublicResults/Sig_L1_TAU12I_eff_BDT_tau_pt.pdf
19. ATLAS Collaboration, G. Aad et al., Identification and energy calibration of hadronically decaying tau leptons with the ATLAS experiment in pp collisions at $\sqrt{s} = 8$ TeV. Eur. Phys. J. **C75**(7), 303 (2015). https://doi.org/10.1140/epjc/s10052-015-3500-z, arXiv:1412.7086 [hep-ex]
20. N. Konstantinidis, L1 track trigger for ATLAS. Technical Report, ATL-UPGRADE-PROC-2010-003, CERN, Geneva, Feb 2010, https://cds.cern.ch/record/1237409
21. ATLAS Collaboration, Early inner detector tracking performance in the 2015 data at $\sqrt{s} = 13$ TeV. Technical Report, ATL-PHYS-PUB-2015-051, CERN, Geneva, Dec 2015, https://cds.cern.ch/record/2110140
22. ATLAS Collaboration, Performance of the ATLAS inner detector track and vertex reconstruction in the high pile-up LHC environment. Technical Report, ATLAS-CONF-2012-042, CERN, Geneva, Mar 2012, https://cds.cern.ch/record/1435196
23. ATLAS Collaboration, G. Aad et al., Measurement of the muon reconstruction performance of the ATLAS detector using 2011 and 2012 LHC proton-proton collision data. Eur. Phys. J. **C74**(11), 3130 (2014). https://doi.org/10.1140/epjc/s10052-014-3130-x, arXiv:1407.3935 [hep-ex]
24. ATLAS Collaboration, G. Aad et al., Muon reconstruction performance of the ATLAS detector in proton-proton collision data at $\sqrt{s} = 13$ TeV. Eur. Phys. J. **C76**(5), 292 (2016). https://doi.org/10.1140/epjc/s10052-016-4120-y, arXiv:1603.05598 [hep-ex]
25. ATLAS Collaboration, G. Aad et al., Electron and photon energy calibration with the ATLAS detector using LHC Run 1 data. Eur. Phys. J. **C74**(10), 3071 (2014). https://doi.org/10.1140/epjc/s10052-014-3071-4, arXiv:1407.5063 [hep-ex]
26. ATLAS Collaboration, Electron efficiency measurements with the ATLAS detector using the 2012 LHC proton-proton collision data. Technical Report, ATLAS-CONF-2014-032, CERN, Geneva, Jun 2014, http://cds.cern.ch/record/1706245
27. ATLAS Collaboration, Electron efficiency measurements with the ATLAS detector using the 2015 LHC proton-proton collision data. Technical Report, ATLAS-CONF-2016-024, CERN, Geneva, Jun 2016, http://cds.cern.ch/record/2157687
28. ATLAS Collaboration, M. Aaboud et al., Measurement of the photon identification efficiencies with the ATLAS detector using LHC Run-1 data, arXiv:1606.01813 [hep-ex]
29. ATLAS Collaboration, Photon identification in 2015 ATLAS data. Technical Report, ATL-PHYS-PUB-2016-014, CERN, Geneva, Aug 2016, http://cds.cern.ch/record/2203125
30. M. Cacciari, G.P. Salam, G. Soyez, The Anti-k(t) jet clustering algorithm. JHEP **04**, 063 (2008). https://doi.org/10.1088/1126-6708/2008/04/063, arXiv:0802.1189 [hep-ph]
31. ATLAS Collaboration, Data-driven determination of the energy scale and resolution of jets reconstructed in the ATLAS calorimeters using dijet and multijet events at $\sqrt{s} = 8$ TeV. Technical Report, ATLAS-CONF-2015-017, CERN, Geneva, Apr 2015, http://cds.cern.ch/record/2008678
32. ATLAS Collaboration, Jet calibration and systematic uncertainties for jets reconstructed in the ATLAS detector at $\sqrt{s} = 13$ TeV. Technical Report, ATL-PHYS-PUB-2015-015, CERN, Geneva, Jul 2015, http://cds.cern.ch/record/2037613

33. ATLAS Collaboration, G. Aad et al., Performance of b-Jet identification in the ATLAS experiment. JINST **11**(04), P04008 (2016). https://doi.org/10.1088/1748-0221/11/04/P04008, arXiv:1512.01094 [hep-ex]
34. ATLAS Collaboration, Optimisation of the ATLAS b-tagging performance for the 2016 LHC run. Technical Report, ATL-PHYS-PUB-2016-012, CERN, Geneva, Jun 2016, http://cds.cern.ch/record/2160731
35. ATLAS Collaboration, Commissioning of the reconstruction of hadronic tau lepton decays in ATLAS using pp collisions at $\sqrt{s}=13$ TeV. Technical Report, ATL-PHYS-PUB-2015-025, CERN, Geneva, Jul 2015, http://cds.cern.ch/record/2037716
36. ATLAS Collaboration, Reconstruction, energy calibration, and identification of hadronically decaying tau leptons in the ATLAS experiment for Run-2 of the LHC. Technical Report, ATL-PHYS-PUB-2015-045, CERN, Geneva, Nov 2015, http://cds.cern.ch/record/2064383
37. ATLAS Collaboration, G. Aad et al., Performance of algorithms that reconstruct missing transverse momentum in $\sqrt{s}=8$ TeV proton-proton collisions in the ATLAS detector, arXiv:1609.09324 [hep-ex]
38. ATLAS Collaboration, Performance of missing transverse momentum reconstruction for the ATLAS detector in the first proton-proton collisions at at $\sqrt{s}=13$ TeV. Technical Report, ATL-PHYS-PUB-2015-027, CERN, Geneva, Jul 2015, http://cds.cern.ch/record/2037904
39. A. Buckley et al., General-purpose event generators for LHC physics. Phys. Rept. **504**, 145–233 (2011). https://doi.org/10.1016/j.physrep.2011.03.005, arXiv:1101.2599 [hep-ph]
40. G. Altarelli, G. Parisi, Asymptotic freedom in parton language. Nucl. Phys. B **126**, 298–318 (1977). https://doi.org/10.1016/0550-3213(77)90384-4
41. Y.L. Dokshitzer, Calculation of the structure functions for deep inelastic scattering and e^+e^- Annihilation by perturbation theory in quantum chromodynamics. Sov. Phys. JETP **46**, 641–653 (1977)
42. V.N. Gribov, L.N. Lipatov, Deep inelastic e p scattering in perturbation theory. Sov. J. Nucl. Phys. **15**, 438–450 (1972)
43. B. Andersson, G. Gustafson, G. Ingelman, T. Sjostrand, Parton fragmentation and string dynamics. Phys. Rep. **97**, 31–145 (1983). https://doi.org/10.1016/0370-1573(83)90080-7
44. F. Krauss, Matrix elements and parton showers in hadronic interactions. JHEP **08**, 015 (2002). https://doi.org/10.1088/1126-6708/2002/08/015, arXiv:hep-ph/0205283 [hep-ph]
45. J. Alwall, M. Herquet, F. Maltoni, O. Mattelaer, T. Stelzer, MadGraph 5: going beyond. JHEP **06**, 128 (2011). https://doi.org/10.1007/JHEP06(2011)128, arXiv:1106.0522 [hep-ph]
46. J. Alwall, R. Frederix, S. Frixione, V. Hirschi, F. Maltoni, O. Mattelaer, H.S. Shao, T. Stelzer, P. Torrielli, M. Zaro, The automated computation of tree-level and next-to-leading order differential cross sections, and their matching to parton shower simulations. JHEP **1407**, 079 (2014). https://doi.org/10.1007/JHEP07(2014)079, arXiv:1405.0301 [hep-ph]
47. T. Gleisberg, S. Hoeche, F. Krauss, M. Schonherr, S. Schumann, F. Siegert, J. Winter, Event generation with SHERPA 1.1. JHEP **02**, 007 (2009). https://doi.org/10.1088/1126-6708/2009/02/007, arXiv:0811.4622 [hep-ph]
48. T. Sjostrand, S. Mrenna, P.Z. Skands, PYTHIA 6.4 physics and manual. JHEP **05**, 026 (2006). https://doi.org/10.1088/1126-6708/2006/05/026, arXiv:hep-ph/0603175 [hep-ph]
49. T. Sjöstrand, S. Ask, J.R. Christiansen, R. Corke, N. Desai, P. Ilten, S. Mrenna, S. Prestel, C.O. Rasmussen, P.Z. Skands, An Introduction to PYTHIA 8.2. Comput. Phys. Commun. **191**, 159–177 (2015). https://doi.org/10.1016/j.cpc.2015.01.024, arXiv:1410.3012 [hep-ph]
50. P. Skands, S. Carrazza, J. Rojo, Tuning PYTHIA 8.1: the Monash 2013 Tune. Eur. Phys. J. **C74**(8), 3024 (2014). https://doi.org/10.1140/epjc/s10052-014-3024-y, arXiv:1404.5630 [hep-ph]
51. M. Bahr et al., Herwig++ physics and manual. Eur. Phys. J. **C58**, 639–707 (2008). https://doi.org/10.1140/epjc/s10052-008-0798-9, arXiv:0803.0883 [hep-ph]
52. J.M. Butterworth, J.R. Forshaw, M.H. Seymour, Multiparton interactions in photoproduction at HERA. Z. Phys. **C72**, 637–646 (1996). https://doi.org/10.1007/BF02909195, https://doi.org/10.1007/s002880050286, arXiv:hep-ph/9601371 [hep-ph]
53. S. Frixione, B.R. Webber, The MC@NLO 3.2 event generator, arXiv:hep-ph/0601192 [hep-ph]

References

54. B.P. Kersevan, E. Richter-Was, The Monte Carlo event generator AcerMC version 1.0 with interfaces to PYTHIA 6.2 and HERWIG 6.3. Comput. Phys. Commun. **149**, 142–194 (2003). https://doi.org/10.1016/S0010-4655(02)00592-1, arXiv:hep-ph/0201302 [hep-ph]
55. M. Mangano et al., Alpgen, a generator for hard multiparton processes in hadronic collisions. JHEP **07**, 001 (2003), arXiv:hep-ph/0206293
56. S. Alioli, P. Nason, C. Oleari, E. Re, A general framework for implementing NLO calculations in shower Monte Carlo programs: the POWHEG BOX. JHEP **1006**, 043 (2010). https://doi.org/10.1007/JHEP06(2010)043, arXiv:1002.2581 [hep-ph]
57. ATLAS Collaboration, G. Aad et al., The ATLAS simulation infrastructure. Eur. Phys. J. **C70**, 823–874 (2010). https://doi.org/10.1140/epjc/s10052-010-1429-9, arXiv:1005.4568 [physics.ins-det]
58. GEANT4 Collaboration, S. Agostinelli et al., GEANT4: a simulation toolkit. Nucl. Instrum. Meth. **A506**, 250–303 (2003). https://doi.org/10.1016/S0168-9002(03)01368-8
59. J.D. Chapman, K. Assamagan, P. Calafiura, D. Chakraborty, D. Costanzo, A. Dell'Acqua, A.D. Simone, G. Lima, Z. Marshall, A. Rimoldi, I. Ueda, S. Vahsen, D. Wright, Y. Zhou, The ATLAS detector digitization project for 2009 data taking. J. Phys. Conf. Ser. **219**(3), 032031 (2010), http://stacks.iop.org/1742-6596/219/i=3/a=032031
60. ATLAS Collaboration, M. Beckingham, M. Duehrssen, E. Schmidt, M. Shapiro, M. Venturi, J. Virzi, I. Vivarelli, M. Werner, S. Yamamoto, T. Yamanaka, The simulation principle and performance of the ATLAS fast calorimeter simulation FastCaloSim. Technical Report, ATL-PHYS-PUB-2010-013, CERN, Geneva, Oct 2010, http://cds.cern.ch/record/1300517

Chapter 6
Validity of Effective Field Theory Dark Matter Models at the LHC

There are different approaches to search for Dark Matter (DM) at the LHC. One strategy is to assume theoretically well-motivated complete models that provide a DM candidate, such as SUSY (introduced in Chap. 4) or extra dimensions. The searches are optimised to target specific decay scenarios and final states that are expected for these models. The characteristics of these final states might depend on specific choices made for the model parameters. Most importantly, the interpretation of the experimental results and the conclusions drawn from them strongly depend on the specific model and its details. This model dependence might be reduced by the use of *Simplified Models* that capture only parts of the model characteristics, as in the case for the *Stop Analsysis* that is presented in Chap. 9. The price to pay for more generality is always a loss of completeness.

The strategy that is widely followed for DM searches at the LHC is the opposite: the most general final state is assumed and the interpretation tries to be as ignorant as possible about the details of any full theory behind the phenomenology that is tested. The Effective Field Theory (EFT) approach plays a key role here. By parametrising everything in the theory apart from the DM particles and their properties an interpretation as model-independent as possible should be obtained. However, the necessary loss of completeness of such an approach shows to be problematic for LHC searches (see e.g. Refs. [1–4]), as will be detailed in the following chapter.

The work presented here started with informal discussions at the University of Geneva between Andrea de Simone, Antonio Riotto, Xin Wu and myself, aiming for bringing together the experimental and the theoretical perspectives on EFT interpretations of LHC DM searches. It resulted in a publication [5] which presented one of the first quantitative assessments of the impact of limited validity of EFTs on the DM search interpretations. The described rescaling procedure was thereafter adapted by the experimental collaborations and the conclusion that an extension of EFT models towards Simplified Models is needed was followed up.

After a discussion of several conditions for the validity of EFTs in Sect. 6.1, the analytical approach including the main observable, R_Λ^{tot} are presented in Sect. 6.2.

The results are then compared to the findings of a numerical analysis, for which I was responsible, in Sect. 6.3. Finally, the results and their impact on the bounds from experiments are discussed in Sect. 6.4 and concluded in Sect. 6.5.

6.1 Effective Field Theory Models of Dark Matter and Their Validity

Based on the general considerations and the set of effective operators discussed in Sect. 3.6, several considerations concerning the validity of an EFT model can be formulated. Roughly speaking, the cut-off scale Λ is the scale where the EFT is expected to break down; however, results can already substantially deviate from those of a full theory at scales below the cut-off.

In the following, a simple model is considered in which a heavy mediator of mass M_{med} is coupled to quarks and DM with couplings g_q and g_χ, respectively. The cut-off scale Λ is assumed to be connected to the mediator mass and couplings via:

$$\Lambda = \frac{M_{\text{med}}}{g_q g_\chi}. \tag{6.1}$$

The EFT is considered being reliable if $Q_{\text{tr}} < M_{\text{med}}$. This, together with the condition of perturbativity of the couplings $g_{q,\chi} < 4\pi$, implies:

$$\Lambda > \frac{Q_{\text{tr}}}{\sqrt{g_q g_\chi}} > \frac{Q_{\text{tr}}}{4\pi}. \tag{6.2}$$

If the momentum transfer is assumed to occur in the s-channel, then kinematics impose $Q_{\text{tr}} > 2m_{\text{DM}}$ and Eq. 6.2 becomes:

$$\Lambda > \frac{m_{\text{DM}}}{2\pi}. \tag{6.3}$$

This requirement is minimal and is refined on an event-by-event basis by the stronger condition of Eq. 6.2, which depends on m_{DM} through Q_{tr}. It is clear that the exact form of a condition like Eq. 6.2 depends on the values of the couplings in the complete theory and its detailed structure. For the following considerations, the couplings g_q and g_χ are assumed to be of order one. Hence, the mediator mass M_{med} can be identified with the suppression scale of the EFT operators, Λ. The above condition then changes to

$$Q_{\text{tr}} < \Lambda. \tag{6.4}$$

However, typical limits on the cut-off scale of an EFT, coming from LHC analyses, are generally below one TeV. It becomes questionable whether an EFT approach is adequate, given that momentum transfers above this scale can well occur in collisions at the LHC.

6.2 Analytical Analysis of the Effective Field Theory Validity

In the following, s-channel exchange of a heavy mediator is assumed. For example, the D1' (D5) operators discussed in Sect. 3.6 correspond to a tree-level s-channel exchange of a heavy scalar (vector) boson S (V_μ), with Lagrangians:

$$\mathcal{L}_{D1'} \supset \frac{1}{2} M^2 S^2 - g_q \bar{q} q S - g_\chi \bar{\chi} \chi S, \tag{6.5}$$

$$\mathcal{L}_{D5} \supset \frac{1}{2} M^2 V^\mu V_\mu - g_q \bar{q} \gamma^\mu q V_\mu - g_\chi \bar{\chi} \gamma^\mu \chi V_\mu. \tag{6.6}$$

The tree-level differential cross sections for the hard scattering process of two incoming particles f and $\bar{f}$ going to two Dark Matter particles χ, with a gluon radiated from the initial state with momentum k, are computed:

$$f(p_1) + \bar{f}(p_2) \to \chi(p_3) + \chi(p_4) + g(k), \tag{6.7}$$

where f can be either a quark (for operators D1–D10), or a gluon (for operators D11–D14). The results are formulated in terms of the momentum transfer in the s-channel:

$$Q_{\mathrm{tr}}^2 = (p_1 + p_2 - k)^2 = x_1 x_2 s - \sqrt{s}\, p_T \left(x_1 e^{-\eta} + x_2 e^\eta\right), \tag{6.8}$$

where x_1, x_2 are the fractions of the proton moment carried by the initial partons, $\sqrt{s}$ denotes the centre-of-mass energy of the collision, and η, p_T are the pseudo-rapidity and the transverse momentum of the final state gluon, respectively. In order to calculate the cross sections, the convolution with the PDFs of the colliding protons needs to be taken. For example, for processes with initial state quarks the following expression needs to be considered for each operator Di:

$$\left.\frac{d^2\sigma}{dp_T d\eta}\right|_{Di} = \sum_q \int dx_1 dx_2 [f_q(x_1) f_{\bar{q}}(x_2) + f_q(x_2) f_{\bar{q}}(x_1)] \left.\frac{d^2\hat{\sigma}}{dp_T d\eta}\right|_{Di}. \tag{6.9}$$

The analytical calculation is performed only for the emission of a gluon from the initial partons, leading to the final state jet. The smaller contribution, coming from initial radiation of quarks ($qg \to \chi\chi + q$),[1] is included in the numerical results presented in Sect. 6.3. The found expressions are valid for all possible values of the parameters. A dependence on η and p_T of the parton is introduced when integrating numerically over the PDFs. The resulting cross sections are:

---

[1] The radiation of quark jets requires a gluon in the initial state. Following the proton PDFs, the gluonic parton momentum fractions are generally below the ones of the quarks. Since the considered final states require relatively high momenta, quarks in the initial state are preferred and hence quark radiation from an initial-state gluon is disfavoured.

$$\frac{d^2\hat{\sigma}}{dp_T d\eta}\bigg|_{D1'} = \frac{\alpha_s}{36\pi^2} \frac{1}{p_T} \frac{1}{\Lambda^4} \frac{[Q_{tr}^2 - 4m_{DM}^2]^{3/2}\left[1 + \frac{Q_{tr}^4}{(x_1 x_2 s)^2}\right]}{Q_{tr}}, \tag{6.10}$$

$$\frac{d^2\hat{\sigma}}{dp_T d\eta}\bigg|_{D4'} = \frac{\alpha_s}{36\pi^2} \frac{1}{p_T} \frac{1}{\Lambda^4} Q_{tr} [Q_{tr}^2 - 4m_{DM}^2]^{1/2}\left[1 + \frac{Q_{tr}^4}{(x_1 x_2 s)^2}\right], \tag{6.11}$$

$$\frac{d^2\hat{\sigma}}{dp_T d\eta}\bigg|_{D5} = \frac{\alpha_s}{27\pi^2} \frac{1}{p_T} \frac{1}{\Lambda^4} \frac{[Q_{tr}^2 - 4m_{DM}^2]^{1/2}[Q_{tr}^2 + 2m_{DM}^2]\left[1 + \frac{Q_{tr}^4}{(x_1 x_2 s)^2} - 2\frac{p_T^2}{x_1 x_2 s}\right]}{Q_{tr}}, \tag{6.12}$$

$$\frac{d^2\hat{\sigma}}{dp_T d\eta}\bigg|_{D8} = \frac{\alpha_s}{27\pi^2} \frac{1}{p_T} \frac{1}{\Lambda^4} \frac{[Q_{tr}^2 - 4m_{DM}^2]^{3/2}\left[1 + \frac{Q_{tr}^4}{(x_1 x_2 s)^2} - 2\frac{p_T^2}{x_1 x_2 s}\right]}{Q_{tr}}, \tag{6.13}$$

$$\frac{d^2\hat{\sigma}}{dp_T d\eta}\bigg|_{D9} = \frac{2\alpha_s}{27\pi^2} \frac{1}{p_T} \frac{1}{\Lambda^4} \frac{\sqrt{Q_{tr} - 4m_{DM}^2}[Q_{tr}^2 + 2m_{DM}^2]\left[1 + \frac{Q_{tr}^4}{(x_1 x_2 s)^2} + 4p_T^2\left(\frac{1}{Q_{tr}^2} - \frac{1}{x_1 x_2 s}\right)\right]}{Q_{tr}}, \tag{6.14}$$

$$\frac{d^2\hat{\sigma}}{dp_T d\eta}\bigg|_{D11} = \frac{3\alpha_s^3}{256\pi^2 \Lambda^6} \frac{(x_1 x_2 s)^3}{(Q_{tr}^2 - x_1 x_2 s)^2} \frac{(Q_{tr}^2 - 4m_{DM}^2)^{3/2}}{p_T Q_{tr}} \left[1 - 4\frac{Q_{tr}^2 - p_T^2}{x_1 x_2 s}\right.$$
$$+ \frac{8Q_{tr}^4 + 21p_T^4}{(x_1 x_2 s)^2} - 2Q_{tr}^2 \frac{5Q_{tr}^4 + 4Q_{tr}^2 p_T^2 + 5p_T^4}{(x_1 x_2 s)^3} + Q_{tr}^4 \frac{8Q_{tr}^4 + 8Q_{tr}^2 p_T^2 + 5p_T^4}{(x_1 x_2 s)^4}$$
$$\left. - 4Q_{tr}^8 \frac{Q_{tr}^2 + p_T^2}{(x_1 x_2 s)^5} + \frac{Q_{tr}^{12}}{(x_1 x_2 s)^6}\right]. \tag{6.15}$$

Some operators are found to be identical to one of the presented operators in the limit of massless light quarks:

$$\frac{d^2\hat{\sigma}}{dp_T d\eta}\bigg|_{D2'} = \frac{d^2\hat{\sigma}}{dp_T d\eta}\bigg|_{D4'} \qquad \frac{d^2\hat{\sigma}}{dp_T d\eta}\bigg|_{D3'} = \frac{d^2\hat{\sigma}}{dp_T d\eta}\bigg|_{D1'}$$
$$\frac{d^2\hat{\sigma}}{dp_T d\eta}\bigg|_{D6} = \frac{d^2\hat{\sigma}}{dp_T d\eta}\bigg|_{D8} \qquad \frac{d^2\hat{\sigma}}{dp_T d\eta}\bigg|_{D7} = \frac{d^2\hat{\sigma}}{dp_T d\eta}\bigg|_{D5}$$
$$\frac{d^2\hat{\sigma}}{dp_T d\eta}\bigg|_{D9} = \frac{d^2\hat{\sigma}}{dp_T d\eta}\bigg|_{D10} \tag{6.16}$$

The results for other operators and details of the derivation of Eqs. (6.10)–(6.15) can be found in Appendix B. The cross sections for the high-energy completions of the dimension-6 operators, with s-channel exchange of a mediator of mass M_{med}, are obtained by the replacement $1/\Lambda^4 \to g_q^2 g_\chi^2/[Q_{tr}^2 - M_{\text{med}}^2]^2$.[2]

---

[2] This relation takes a slightly more complicated form for some of the operators, which is not considered here.

6.2.1 Quantifying the Effective Field Theory Validity

As discussed above, the truncation to the lowest-dimensional operator of the EFT expansion is applicable only if the momentum transfer is smaller than an energy scale of the order of Λ (see Eq. 6.4). The fraction of events with momentum transfer lower than the EFT cut-off scale is considered as a measure of validity. The ratio of the EFT cross section obtained with imposing $Q_{tr} < \Lambda$, over the total EFT cross section is defined to this end:

$$R_\Lambda^{tot} \equiv \frac{\sigma|_{Q_{tr}<\Lambda}}{\sigma} = \frac{\int_{p_T^{min}}^{p_T^{max}} dp_T \int_{\eta_{min}}^{\eta_{max}} d\eta \left. \frac{d^2\sigma}{dp_T d\eta}\right|_{Q_{tr}<\Lambda}}{\int_{p_T^{min}}^{p_T^{max}} dp_T \int_{\eta_{min}}^{\eta_{max}} d\eta \frac{d^2\sigma}{dp_T d\eta}}. \quad (6.17)$$

To sum over the possible p_T and η of the jets, the differential cross sections are integrated over values typically considered in experimental searches. In the following, $p_T^{min} = 500$ GeV (as used in the signal region SR4 of Ref. [6]) and $|\eta| < 2$ are assumed for centre-of-mass energies of $\sqrt{s} = 8$ and 14 TeV. For p_T^{max}, values of 1 (2) TeV were used for $\sqrt{s} = 8(14)$ TeV, respectively. The sum over quark flavours is performed considering only u, d, c, s quarks.

First, the behaviour of the ratio R_Λ^{tot}, as a function of Λ and m_{DM} is studied. The results for the operators $D1'$, $D5$, $D9$ are shown in Fig. 6.1. The ratio R_Λ^{tot} approaches unity for large values of Λ, as in this case the effect of the cut-off becomes negligible and the EFT can be considered fully applicable. The ratio decreases for large m_{DM}, because the momentum transfer is necessarily higher in this regime in order to allow for the production of the heavier DM particles. Going from $\sqrt{s} = 8$ TeV to $\sqrt{s} = 14$ TeV, the results scale almost linearly with the energy: for the same value of the ratio of m_{DM} over Λ nearly the same R_Λ^{tot} is obtained.

The contours of constant values of the quantity R_Λ^{tot} can be defined in the plane (m_{DM}, Λ). These curves are shown in Fig. 6.2 for different operators for $\sqrt{s} = 8$ TeV and in Fig. 6.3 for $\sqrt{s} = 14$ TeV. For cut-off scales above ~ 1 TeV for $\sqrt{s} = 8$ TeV (~ 2 TeV for $\sqrt{s} = 14$ TeV), in around 50% of the events the momentum transfer is above the cut-off and the EFT should be considered as invalid.

The contours for $D1'$–$D4'$ differ from the corresponding contours for $D1$–$D4$ by factors of order one, due to the different weighting of the quark PDFs. The experimental limits on the EFT cut-off scale for these operators are only of the order of tens of GeV, as their cross section experiences an additional suppression of m_q/Λ. In the experimentally probed region, the EFT for interactions of the type of the operators $D1$–$D4$ is far from being valid and hence the results are not reliable.

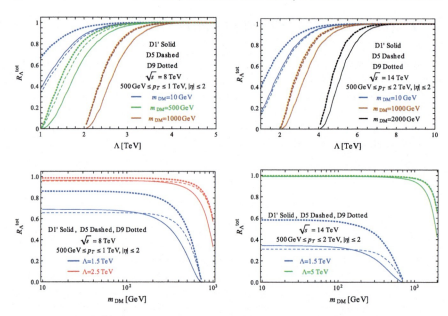

Fig. 6.1 The ratio R_Λ^{tot} defined in Eq. (6.17) for operators $D1'$ (solid lines), $D5$ (dashed lines) and $D9$ (dotted lines) as a function of Λ (top) and m_{DM} (bottom), for $\sqrt{s} = 8$ TeV (left) and 14 TeV (right)

As mentioned, the precise connection between the EFT cut-off scale and the mediator mass depends on the details of the unknown high-energy completion of the model. When the couplings of the complete theory reach their maximal values allowed by perturbativity, the validity requirement becomes $Q_{\text{tr}} < 4\pi\Lambda$. This presents the most optimistic scenario in view of the EFT validity. The effect of varying the assumed couplings is shown in Fig. 6.4 for the contour $R_\Lambda^{\text{tot}} = 50\%$ of the $D5$ operator. Other operators show similar results, as the contours scale linearly with the cut-off.

For comparison, a grey shaded area indicating the region where $\Lambda < m_{\text{DM}}/(2\pi)$ is shown. This bound is often quoted as an indication for the non-validity of the EFT (see Eq. 6.3). The 50% contour of R_Λ^{tot} is above such a region, meaning that issues of validity of the EFT approach need to be considered even when this condition is fulfilled.

In conclusion, interpreting the experimental data in terms of an EFT can lead to significantly different results than if a mediator would be included in the model, if Λ is found to be of the order of a few TeV.

6.2 Analytical Analysis of the Effective Field Theory Validity

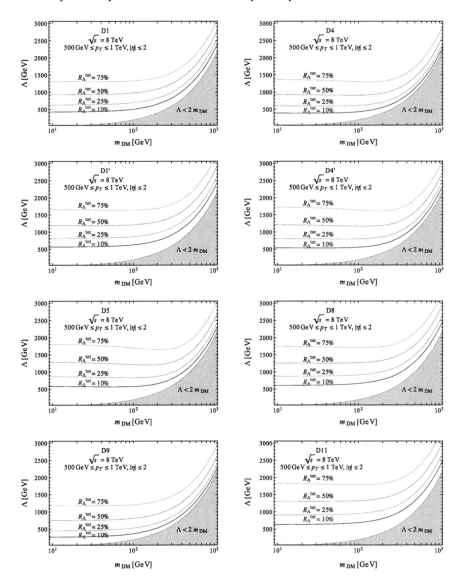

Fig. 6.2 Contours for ratios R_Λ^{tot}, defined in Eq. (6.17), of 10–75% in the plane (m_{DM}, Λ), for several operators at $\sqrt{s} = 8$ TeV. The grey area corresponds to the region in which the EFT assumption is definitely invalid, since $\Lambda < 2m_{DM}$

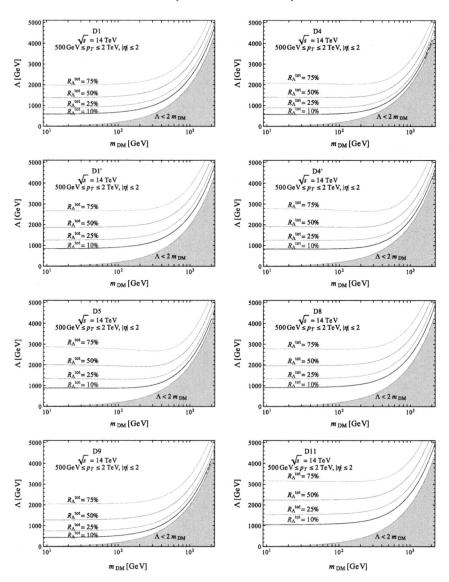

Fig. 6.3 Contours for ratios R_Λ^{tot}, defined in Eq. (6.17), of 10–75% in the plane (m_{DM}, Λ), for several operators at $\sqrt{s} = 14$ TeV. The grey area corresponds to the region in which the EFT assumption is definitely invalid, since $\Lambda < 2m_{DM}$

6.3 Numerical Approach to Effective Field Theory Validity

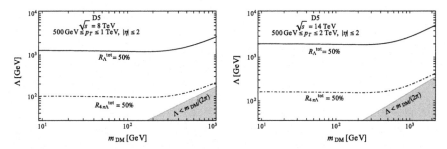

Fig. 6.4 The 50% contours for the ratio R_Λ^{tot} for the operator $D5$, varying the coupling choice from one to 4π and hence the condition on the momentum transfer from $Q_{\text{tr}} < \Lambda$ (solid line) to $Q_{\text{tr}} < 4\pi\Lambda$ (dot-dashed line). Also shown is the region corresponding to $\Lambda < m_{\text{DM}}/(2\pi)$ (grey shaded area). Both collision energies, $\sqrt{s} = 8$ TeV (left panel) and $\sqrt{s} = 14$ TeV (right panel), are considered

6.3 Numerical Approach to Effective Field Theory Validity

In order to verify and study further the analytical findings and to better compare to the experimental limits, results on the ratio R_Λ^{tot} are also computed using a Monte Carlo event generator.

6.3.1 Simulation and Analysis Description

For the simulation of pp collisions at $\sqrt{s} = 8$ TeV and $\sqrt{s} = 14$ TeV, MADGRAPH 5 [7] is used. Both PDF sets CTEQ6L1 and MSTW2008LO (discussed in Ref. [8]) are employed. The PDF choice affects the cross section, but has only minimal effects on the acceptance. Hence, the change in contours of R_Λ^{tot} is negligible. Since MSTW2008LO is used for the analytical calculations, this PDF set is applied where direct comparisons between simulation and calculation are shown. For the comparison with the experimental results, CTEQ6L1 is used instead, as was done in their interpretation. Only u, d, c, s quarks were considered, both in the initial and in the final state.

From the event kinematics, it is evaluated whether or not the conditions of validity discussed in Sect. 6.2 are fulfilled. Specifically, it was checked, if Eq. 6.2 holds, that is, if the following condition is met:

$$\Lambda > \frac{Q_{\text{tr}}}{\sqrt{g_q g_\chi}} > 2\frac{m_{\text{DM}}}{\sqrt{g_q g_\chi}}. \tag{6.18}$$

Samples of 20,000 events are simulated for each operator, scanning DM mass values of 10, 50, 80, 100, 400, 600, 800 and 1000 GeV and cut-off scales of 250, 500, 1000, 1500, 2000, 2500 and 3000 GeV in the case of $\sqrt{s} = 8$ TeV collisions. When increasing the collision energy to $\sqrt{s} = 14$ TeV, the DM mass of 2000 GeV and cut-off scales of 4000 and 5000 GeV are added.

From the simulated samples, the fraction of events fulfilling $\Lambda > Q_{tr}/\sqrt{g_q g_\chi}$ can be evaluated for each pair of DM mass and cut-off scale, if a certain value for the couplings $\sqrt{g_\chi g_q}$ connecting the cut-off scale Λ and the mediator mass M via $\Lambda = M/\sqrt{g_q g_\chi}$ is assumed. As in the analytical approach, $g_q g_\chi$ is assumed to be one.

6.3.2 Results

In order to confirm that analytical and numerical results are in agreement, Fig. 6.5 shows a comparison for the operators $D1'$, $D4'$, $D5$, $D8$ and $D9$. The results were obtained for the scenario of one radiated gluon jet above 500 GeV within $|\eta| < 2$. The contours of $R_\Lambda^{tot} = 50\%$ from analytical and numerical evaluation agree within 7%. The remaining differences are attributed to the upper jet p_T cut not imposed during event simulation but needed for the analytical calculation, and the details of the respective fitting procedures applied to extract the percentiles of R_Λ^{tot}.

The kinematic constraints are varied step by step from the scenario considered in the analytical calculations, namely one radiated gluon jet above 500 GeV within $|\eta| < 2$, to a scenario closest to the analysis cuts applied in the ATLAS *Monojet Analysis* [6]. More specifically, the leading jet is allowed to come from either a gluon or a quark being radiated, the leading jet p_T cut is lowered from 500 to 350 GeV, a second jet is allowed and its range in η is enlarged to $|\eta| < 4.5$. No further cuts are applied at simulation level. The effect of the variation of the cuts can be seen in

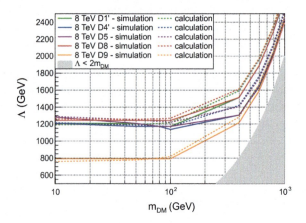

Fig. 6.5 Comparison of the contour $R_\Lambda^{tot} = 50\%$ for the analytical calculation (dashed line) and the simulation (solid line) for the different operators $D1'$, $D4'$, $D5$, $D8$ and $D9$. The results agree within 7%. The grey area corresponds to the region in which the EFT assumption is definitely invalid, since $\Lambda < 2m_{DM}$

6.3 Numerical Approach to Effective Field Theory Validity

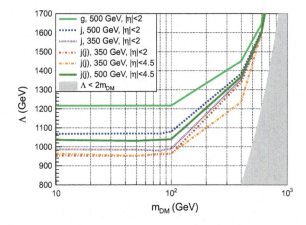

Fig. 6.6 Changes of the contour of $R_\Lambda^{\text{tot}} = 50\%$ are shown for several variations from a scenario close to the analytical calculation to one close to the cuts used in the ATLAS monojet analysis [6], exemplarily for the operator $D5$ at $\sqrt{s} = 8$ TeV. In the legend, "g" means that only gluon radiation is considered, "j" stands for either quark- or gluon-initiated jets, "j(j)" means that a second jet is allowed to appear. The grey area corresponds to the region in which the EFT assumption is definitely invalid, since $\Lambda < 2m_{DM}$

Fig. 6.6. Allowing not only for a gluon jet but also taking into account the possibility of a quark jet changes the R_Λ^{tot} contours appreciably. The change from lowering the p_T of the leading jet has a smaller effect. Allowing for a second jet and enhancing its rapidity range barely changes the R_Λ^{tot} contour, especially at large m_{DM} values. If the collision energy is augmented to $\sqrt{s} = 14$ TeV, all the R_Λ^{tot} contours increase. As for $\sqrt{s} = 8$ TeV, moving to the scenario closer to the experimental analysis leads to contours that are at most 30% lower in Λ.

After having extracted R_Λ^{tot} for each DM mass and cut-off scale, a curve can be fitted through the points obtained in the plane of R_Λ^{tot} and Λ to extract the percentiles of R_Λ^{tot}. The following functional form is used for this purpose:

$$R_\Lambda^{\text{tot}} = \left[1 - e^{-a\left(\frac{\Lambda - 2m_{DM}}{b}\right)^c}\right]\left[1 - e^{-d\left(\frac{\Lambda + 2m_{DM}}{b}\right)^f}\right]. \tag{6.19}$$

The parameters a, b, c, d and f are fitted for each DM mass separately. From these fits, the points denoting a cut-off scale where R_Λ^{tot} equals e.g. 50% can be extracted for each DM mass, and the lines of constant R_Λ^{tot} can be plotted in the usual limit-setting plane (m_{DM}, Λ). Table 6.1 collects the values of the fitted parameters for all operators for which an experimental result is available.

Table 6.1 Fitted parameters of the functions describing R_Λ^{tot} in Eq. (6.19), in the cases of $\sqrt{s} = 8$ and 14 TeV. The fitting functions describe processes where quarks and/or gluons are radiated, the final state contains 1 or 2 jets, where the leading jet has a p_T larger than 350 GeV while the second jet is allowed to be within $|\eta| < 4.5$

Operator	a	b	c	d	e
$\sqrt{s} = 8$ TeV					
D1	1.32	787.13	1.39	1.08	1.53
D1'	1.30	1008.25	1.49	0.77	1.83
D4	1.65	702.93	1.14	0.65	1.75
D4'	1.51	859.83	1.22	0.48	1.92
D5	1.54	816.83	1.18	0.50	1.85
D8	1.23	964.62	1.50	0.91	1.59
D9	1.43	681.92	1.15	1.02	1.35
D11	1.23	1002.33	1.49	0.82	1.69
$\sqrt{s} = 14$ TeV					
D1	0.89	1017.37	1.45	1.28	1.24
D1'	0.43	909.66	1.59	0.53	1.37
D4	1.23	996.82	1.25	0.80	1.48
D4'	0.76	982.75	1.33	0.37	1.63
D5	0.78	894.86	1.25	0.39	1.54
D8	0.48	945.09	1.55	0.74	1.24
D9	0.91	891.65	1.21	1.23	1.04
D11	0.68	1250.49	1.58	0.81	1.35

6.4 Implications on Dark Matter Searches at LHC

Figure 6.7 shows the experimental limits obtained from the ATLAS monojet analysis of 10 fb^{-1} of pp collisions at $\sqrt{s} = 8$ TeV [6] in the plane (Λ, m_{DM}), for the opearators D5, D8 and D11. The percentiles of R_Λ^{tot} of 25, 50 and 75% are superimposed. The experimental limits are placed in a region where about 30% of the events can be expected to fulfil the EFT validity conditions. The exact number depends on the operator considered. Especially the limit on the gluon operator D11 seems questionable in this view. Also shown are the contours of R_Λ^{tot} for couplings of $\sqrt{g_q g_\chi} = 4\pi$, which presents the limiting case in which the theory is still considered perturbative. Since there is no possibility to directly measure Q_{tr} in data, on an event-by-event basis, the information on the fraction of invalid events can only be estimated from analytical calculations or a numerical simulation. The impact of the limited validity of the EFT approach on the current collider bounds is quantified in the following.

It is assumed that imposing EFT validity by the above-mentioned cut on Q_{tr} only changes the normalisation of the distributions of p_T or E_T^{miss} but not their shape.[3]

---

[3] Brief studies indicated that the assumption made here is reasonable for $Q_{tr} < 750$ GeV.

6.4 Implications on Dark Matter Searches at LHC

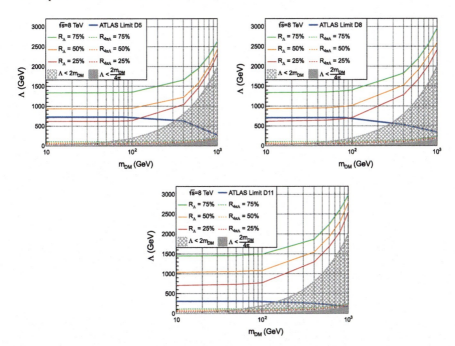

Fig. 6.7 The 25, 50 and 75% contours for the ratio R_Λ^{tot}, compared to the experimental limits from ATLAS [6] (blue line). Also indicated are the contours of R_Λ^{tot} in the extreme case when setting the couplings to $\sqrt{g_q g_\chi} = 4\pi$ (dashed lines). Results are shown for different operators: D5 (upper left panel), D8 (upper right panel) and D11 (lower panel)

Neglecting the statistical and systematic uncertainties, the number of signal events in a given EFT model has to be less than the experimental limit on observed new physics events: $N_{\text{signal}}(\Lambda, m_{\text{DM}}) < N_{\text{obs}}$. The EFT cross section scales like $\Lambda^{-2(d-4)}$ for an operator of mass dimension d. Following this expression, one can write: $N_{\text{signal}}(\Lambda, m_{\text{DM}}) = \Lambda^{-2(d-4)} \tilde{N}_{\text{signal}}(m_{\text{DM}})$, and the experimentally observed lower bound on the scale of the operator becomes

$$\Lambda > \left[\tilde{N}_{\text{signal}}(m_{\text{DM}})/N_{\text{obs}}\right]^{1/[2(d-4)]} \equiv \Lambda_{\text{obs.}}. \qquad (6.20)$$

Some fraction of the considered simulated signal events have a momentum transfer that exceeds the cut-off scale of the EFT. These events are now excluded, i.e. the number of signal events for placing the limit gets reduced by a factor of R_Λ^{tot}. Hence, $N_{\text{signal}}(\Lambda, m_{\text{DM}}) \to R_\Lambda^{\text{tot}}(m_{\text{DM}}) N_{\text{signal}}(\Lambda, m_{\text{DM}})$, and the new limit is determined via:

$$\Lambda > [R_\Lambda^{\text{tot}}(m_{\text{DM}})]^{1/2[(d-4)]} [N_{\text{signal}}(m_{\text{DM}})/N_{\text{obs}}]^{1/[2(d-4)]} = [R_\Lambda^{\text{tot}}(m_{\text{DM}})]^{1/[2(d-4)]} \Lambda_{\text{obs}}, \qquad (6.21)$$

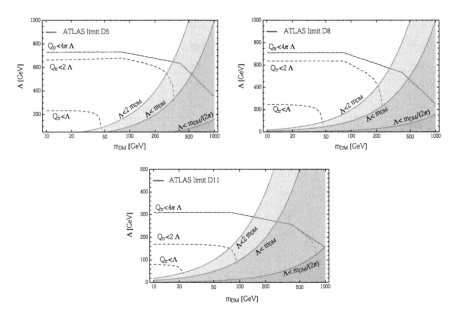

Fig. 6.8 ATLAS experimental limits on the suppression scale Λ [6] are shown as solid blue lines. The rescaled limits, restricting considered signal events to fulfil the EFT validity condition, are shown as dashed black lines, for $Q_{tr} < \Lambda, 2\Lambda, 4\pi\Lambda$, corresponding to different choices of the UV couplings: $\sqrt{g_q g_\chi} = 1, 2, 4\pi$, respectively. The kinematic constraints (Eq. 6.18) are denoted by grey bands. Different operators are shown: $D5$ (upper left panel), $D8$ (upper right panel) and $D11$ (lower panel)

which is weaker than Λ_{obs}.[4] Figure 6.8 shows the recalculated limits for the dimension-6 operators D5 and D8 and the dimension-7 operator D11, under different assumptions on the UV coupling strengths, namely for $\sqrt{g_q g_\chi} = 1, 2, 4\pi$. The fraction of events passing the validity criterion, R_Λ^{tot}, are taken from the numerical results listed in Table 6.1, which include both quark and gluon jets, and a selection close to one of the signal regions ("SR3") used by ATLAS [6]. As expected, the higher the couplings, the weaker the condition on Q_{tr}, the closer the rescaled limits are to the bounds presented by ATLAS. In the limiting case of $\sqrt{g_q g_\chi} = 4\pi$ with the most optimistic condition on the momentum transfer ($Q_{tr} < 4\pi\Lambda$), the new limits for D5 and D8 are indistinguishable from the ATLAS bounds, meaning that the experimental results are safe in terms of EFT validity. Regarding D11, even for extreme values of the couplings, the limit at large DM masses deviates.

---

[4]In principle, this procedure needs to be repeated iteratively, until convergence if achieved: the new bound on Λ replaces Λ_{obs} and, using this new bound, R_Λ^{tot} is re-evaluated, resulting in a new rescaled bound and so on. This is neglected in the present analysis.

6.4 Implications on Dark Matter Searches at LHC

For couplings of order one, the rescaled limits are significantly weaker than those reported. As a result of these studies, experimental collaborations have since adopted such a rescaling procedure and the modified results for EFT models are published along with the nominal limits.

6.5 Conclusions

The search for DM is one of the main targets of the LHC experiments. Investigations into the validity of EFTs commonly used in interpreting such searches have been presented. A measure of the validity of an EFT, R_Λ^{tot}, was introduced. It indicates the fraction of events for which the defined condition of validity for the EFT is fulfilled and depends on the DM mass and the assumed cut-off scale. The analysis for the full list of EFT operators used by the ATLAS and CMS collaborations, connecting fermion DM particles and quarks or gluons and originating from the exchange of heavy mediators in the s-channel, has been performed analytically, assuming collision energies of 8 and 14 TeV. The analytical results were completed by performing numerical event simulations which reproduce the experimental situation as closely as possible. The results indicate that the range of validity of the EFT is significantly limited in the parameter space of (m_{DM}, Λ) that is probed by LHC searches. While these findings are valid for s-channel processes, a similar analysis exists also for t-channel scenarios [9], where comparable results are obtained.

The advantage of avoiding too much model dependence still holds for the EFT approach; however, the presented results clearly demand an alternative to the EFT interpretation, such as through identifying a set of *Simplified Models*, which are able to reproduce the EFT operators in the heavy mediator limit. This allows for a consistent analysis of the current and future LHC data by consistently taking into account the possibility of an on-shell production of the mediator. Furthermore, comparisons to direct and indirect searches can be presented in a more comprehensive way by using such Simplified Models.

In the following chapter, the ATLAS *Monojet Analysis* of 20.3 fb^{-1} of 8 TeV pp collision data is presented, which adapted the rescaling procedure introduced here and presented the EFT limits alongside with these modified bounds. Furthermore, a first step towards the consistent use of Simplified Models is made there by considering a Z'-like model for interpretation. Subsequently, a study is presented that conducted a detailed re-interpretation of this and two other DM searches in terms of a set of Simplified Models and also shows comparisons to direct detection results.

References

1. G. Busoni, A. De Simone, E. Morgante, A. Riotto, On the validity of the effective field theory for dark matter searches at the LHC. Phys. Lett. **B728**, 412–421 (2014). https://doi.org/10.1016/j.physletb.2013.11.069, http://arxiv.org/abs/1307.2253 [hep-ph]
2. O. Buchmueller, M.J. Dolan, C. McCabe, Beyond effective field theory for dark matter searches at the LHC. JHEP **01**, 025 (2014). https://doi.org/10.1007/JHEP01(2014)025, http://arxiv.org/abs/1308.6799 [hep-ph]
3. J. Goodman, M. Ibe, A. Rajaraman, W. Shepherd, T.M.P. Tait, H.-B. Yu, Constraints on dark matter from colliders. Phys. Rev. **D82**, 116010 (2010). https://doi.org/10.1103/PhysRevD.82.116010, http://arxiv.org/abs/1008.1783 [hep-ph]
4. P.J. Fox, C. Williams, Next-to-leading order predictions for dark matter production at hadron colliders. Phys. Rev. **D87**(5), 054030 (2013). https://doi.org/10.1103/PhysRevD.87.054030, http://arxiv.org/abs/1211.6390 [hep-ph]
5. G. Busoni, A. De Simone, J. Gramling, E. Morgante, A. Riotto, On the validity of the effective field theory for dark matter searches at the LHC, Part II: Complete analysis for the s-channel. JCAP **1406**, 060 (2014). https://doi.org/10.1088/1475-7516/2014/06/060, http://arxiv.org/abs/1402.1275 [hep-ph]
6. ATLAS Collaboration, Search for new phenomena in monojet plus missing transverse momentum final states using $10\,\text{fb}^{-1}$ of pp collisions at $\sqrt{s} = 8$ TeV with the ATLAS detector at the LHC. Technical Report, ATLAS-CONF-2012-147, CERN, Geneva, Nov 2012, http://cds.cern.ch/record/1493486
7. J. Alwall, M. Herquet, F. Maltoni, O. Mattelaer, T. Stelzer, MadGraph 5: going beyond. JHEP **06**, 128 (2011). https://doi.org/10.1007/JHEP06(2011)128, http://arxiv.org/abs/1106.0522 [hep-ph]
8. A.D. Martin, W.J. Stirling, R.S. Thorne, G. Watt, Parton distributions for the LHC. Eur. Phys. J. **C63**, 189–285 (2009). https://doi.org/10.1140/epjc/s10052-009-1072-5, http://arxiv.org/abs/0901.0002 [hep-ph]
9. G. Busoni, A. De Simone, T. Jacques, E. Morgante, A. Riotto, On the validity of the effective field theory for dark matter searches at the LHC Part III: Analysis for the t-channel. JCAP **1409**, 022 (2014). https://doi.org/10.1088/1475-7516/2014/09/022, http://arxiv.org/abs/1405.3101 [hep-ph]

Chapter 7
Search for Dark Matter in Monojet-like Events

The simplest possible scenario for the production of Dark Matter (DM) at the LHC is given by assuming a process in which two incoming partons would lead to a final state with two DM particles.[1] However, such a final state could not be detected in the experiments: since the DM particles are only interacting weakly with the detector material, they would escape without leaving a signal.

On the other hand, if the radiation of an object like a jet, a photon or even a vector boson from the initial partons is assumed, the final state presents a very unique "mono-X" signature: one energetic object is the only activity in the event. It is recoiling against the invisible, undetected particles which leads to a significant momentum imbalance in the transverse plane. This scenario is sketched in Fig. 7.1.

Even without having a signal scenario in mind, event topologies featuring an energetic object and large missing transverse energy (E_T^{miss}) are distinct signatures to look for physics beyond the Standard Model at the LHC. Various possible final states have been studied: mono-jet, mono-photon, mono-W/Z and mono-Higgs [1–11]. Due to the large probability of radiating a gluon or a quark off the incoming partons, the mono-jet channel generally has the highest cross section and is most sensitive to possible signals.

Such searches were commonly interpreted in terms of an effective field theory (EFT) of DM production, as introduced in Sect. 3.6. Such models are useful to reduce the model dependence of the interpretation of the results and allow for a straight-forward comparison between collider searches and direct or indirect detection experiments. However, the assumptions entering in the EFT formulation are often not justified at LHC energies, as outlined in Chap. 6. There, a method is introduced with which the obtained EFT limits can be modified, allowing to judge the impact of the limited EFT validity on the resulting limits.

---

[1]A final state featuring only one DM particle might be considered even simpler. However, almost all of the proposed DM models require that these new particles have some kind of conserved charge (e.g. R-parity in SUSY) such that they are stable. This would require them to be produced in pairs.

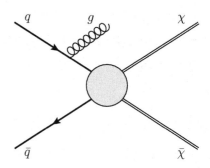

Fig. 7.1 Sketch of the pair production of DM particles, $\chi\bar{\chi}$, associated with a jet from initial-state radiation of a gluon, g

In the following, the search for new physics in events with an energetic jet and large $E_\mathrm{T}^\mathrm{miss}$, performed on the full 8 TeV dataset of 20.3 fb^{-1}, is presented. There are two major improvements with respect to its precursor. First, a veto on isolated tracks is introduced, allowing for a powerful rejection of electroweak backgrounds. Second, a dedicated optimisation for DM signals is performed, leading to the replacement of the veto on additional jets[2] by a topological cut on the balance between the leading jet p_T and the $E_\mathrm{T}^\mathrm{miss}$. The analysis results were published in 2015 [12].

After an introduction to the analysis strategy in Sect. 7.1, and details on the data and simulation samples used (Sect. 7.2), the event selection is introduced in Sect. 7.3. In particular, the veto on isolated tracks, which I developed, is discussed in detail in Sect. 7.3.3, as well as the optimisation for DM signals, to which I contributed (Sect. 7.3.4). The background estimation is explained in Sect. 7.4 and the sources of systematic uncertainties are discussed in Sect. 7.5 before the results are presented in Sect. 7.6. Finally, the resulting model-independent limits on events from new physics and the DM interpretation in terms of EFT and Simplified Models are presented in Sect. 7.7. I played a significant role also in defining the interpretation strategy, the presentation of results, the comparisons to direct and indirect DM searches and I calculated the relic density constraints. Conclusions with an outlook on new results from this channel is given in Sect. 7.8.

7.1 Analysis Strategy

The analysis looks for evidence of new physics in events with one energetic jet and large $E_\mathrm{T}^\mathrm{miss}$. While the occurrence of leptons (electrons and muons) is vetoed, additional jets are allowed under certain conditions. Such events are referred to as "monojet-like" in the following. Nine signal regions (SRs) are defined by an increasing, inclusive lower $E_\mathrm{T}^\mathrm{miss}$ cut.

The analysis faces a large, irreducible background from Z+jets events in which the Z decays to–invisible–neutrinos. To a smaller extend, $Z/\gamma^*(\to \ell^+\ell^-)$ and

---

[2]Events containing more than two jets were vetoed before.

7.1 Analysis Strategy

$W(\to \ell\nu)$ events, in which the lepton(s) are not identified or out of acceptance, contribute to the background of the signal regions. The background contributions from these electroweak processes are estimated from simulation. Their normalisation is extracted from data in background-enriched control regions (CRs). In this approach, so-called transfer factors are used to extrapolate each background from CRs to SRs. This method significantly reduces the impact of systematic uncertainties on the final result.

7.2 Dataset and Simulations

7.2.1 Dataset

The analysis uses the data of LHC pp collisions at $\sqrt{s} = 8$ TeV, recorded with the ATLAS detector during 2012. The mean number of interactions per bunch-crossing (*pile-up*) is 20.7 [13]. Collisions that were recorded during stable beam conditions and fulfilled some basic quality criteria amount to an integrated luminosity of 20.3 fb^{-1}. The uncertainty on this value is estimated to be 2.8% and is derived following the methodology outlined in Ref. [14].

Events that are studied in this search are accepted for recording by a trigger based on E_T^{miss}. The trigger-level reconstruction of E_T^{miss} in this case is entirely based on energy deposits in the calorimeter. As a consequence, muons are not seen by the algorithm and stay "invisible" to the trigger. The input to the algorithm are topologically connected calorimeter clusters (topo-clusters) that are locally calibrated to the hadronic scale. The trigger threshold is $E_T^{\text{miss}} > 80$ GeV. This corresponds to full efficiency at the analysis level for $E_T^{\text{miss}} > 150$ GeV. For all SRs and CRs of the analysis the trigger can be considered fully efficient. An inclusive combination of the single-electron triggers requiring one isolated electron with trigger-level p_T of at least 24 GeV or one electron with trigger-level $p_T > 60$ GeV, without isolation requirements, is used for the $W(\to e\nu)$ and $Z/\gamma^*(\to e^+e^-)$ control samples.

7.2.2 Monte Carlo Simulations

Background Simulation

W/Z + jets: W+jets and Z+jets events are simulated using the SHERPA [15] event generator. The simulation includes leading-order (LO) matrix elements for up to five partons in the final state and assumes massive b/c-quarks. The CT10 [16] parton distribution functions (PDF) of the proton are used. An alternative generator, ALPGEN [17], together with the parton-shower description from HERWIG [18, 19] plus JIMMY [20] and the PDF set CTEQ6L1 [21], was also used to simulate these electroweak processes. These samples have lower statistics and are used for the isolated-track veto optimisation described in Sect. 7.3.3 and the estimation of systematic uncertainties. The calculations are then normalised to next-to-next-to-leading-order (NNLO) perturbative QCD (pQCD) predictions [22] using the MSTW2008 NNLO PDF sets [23].

Top Quark Production: Both top-quark pair production and single-top processes may enter the selection in a small amount. During the event generation, a top-quark mass of 172.5 GeV is assumed. The production of top-quark pairs ($t\bar{t}$), Wt and s-channel single top is simulated using the MC@NLO MC generator [24, 25]. It is interfaced to HERWIG plus JIMMY to model the parton showers and the underlying event. The AcerMC [26] program is used to simulate single-top production in the t-channel. As in the case of W/Z+jets backgrounds, the $t\bar{t}$ [27] and single-top processes [28] are normalised, using the information from NNLO+NNLL (next-to-next-to-leading-logarithm) pQCD cross sections [29].

The AUET2C and AUET2B [30] set of optimised parameters for the underlying event description are used. They rely on the CT10 and CTEQ6L1 [21] PDFs, respectively.

Multijet and γ+jet samples are generated with PYTHIA 8 [31], again using the CT10 PDF.

Dibosons: The diboson sample generation of WW, WZ, ZZ, $W\gamma$ and $Z\gamma$ processes uses SHERPA with the CT10 PDF set, also applying a normalisation from NLO pQCD predictions [32].

Signal Simulation

The signal samples for DM pair production are generated using the MADGRAPH 5 [33] implementation of the effective field theory based model described in Ref. [34]. The effective operators introduced in Sect. 3.6 are considered. They can be grouped according to the expected spectrum of E_T^{miss}. For each of these group, one operator is chosen and simulated: D1 (scalar), D5 (vector), D9 (tensor) and D11 (gluon, scalar) for the assumption of Dirac fermion DM particles, C1 and C5 for complex scalar DM. The considered operators are listed in Table 7.1

The events are generated with one or two jets produced in addition to the DM particles at matrix element level. More jets may be added during the parton showering. The matrix-element description is generally more accurate. The possibly occurring extra jets can have an effect on the signal cross section and acceptance. The number of additional jets produced at matrix element level is optimised: including more jets in the matrix element does not alter the cross section or acceptance appreciably. At least one of these partons is required to have a minimum p_T of 80 GeV. Only initial states of gluons and the four lightest quarks are considered. The coupling strengths between the quarks and the DM particles are assumed to be equal for all quark flavours. Since the operator D1 has an explicit dependence on the quark mass (see

7.2 Dataset and Simulations

Table 7.1 Effective interaction operators of DM interactions with Standard Model quarks or gluons, following the formalism in Ref. [34], where Λ is the EFT cutoff scale. Operators starting with a D assume Dirac fermion DM, the ones starting with a C consider complex scalar DM. $G^a_{\mu\nu}$ is the gluon field-strength tensor

Name	Initial state	Type	Operator
C1	qq	Scalar	$\frac{m_q}{\Lambda^2}\chi^\dagger\chi\bar{q}q$
C5	gg	Scalar	$\frac{1}{4\Lambda^2}\chi^\dagger\chi\alpha_s(G^a_{\mu\nu})^2$
D1	qq	Scalar	$\frac{m_q}{\Lambda^3}\bar{\chi}\chi\bar{q}q$
D5	qq	Vector	$\frac{1}{\Lambda^2}\bar{\chi}\gamma^\mu\chi\bar{q}\gamma_\mu q$
D8	qq	Axial-vector	$\frac{1}{\Lambda^2}\bar{\chi}\gamma^\mu\gamma^5\chi\bar{q}\gamma_\mu\gamma^5 q$
D9	qq	Tensor	$\frac{1}{\Lambda^2}\bar{\chi}\sigma^{\mu\nu}\chi\bar{q}\sigma_{\mu\nu}q$
D11	gg	Scalar	$\frac{1}{4\Lambda^3}\bar{\chi}\chi\alpha_s(G^a_{\mu\nu})^2$

Table 7.1), its cross section is most sensitive to the relevant mass of the charm quark, which is set to 1.42 GeV. Since the fraction of b-flavoured sea-quarks in the colliding protons is non-negligible at the LHC, it would have been preferable to include also b-quarks in possible initial states.

The events generated with MADGRAPH 5 are then processed with PYTHIA 6 [35] for the simulation of the parton showering and hadronisation. The so-called MLM prescription [36] is used to match the matrix-element calculations to the parton shower evolution. A matching scale specifies that MADGRAPH will produce hard jets with momenta above the matching scale, while PYTHIA will control the soft showering and radiation below the matching scale. Such a separation is needed in order not to double-count potential diagrams. Two different matching scales (80 and 300 GeV) are considered to guarantee sufficient statistics in the E_T^miss tail and to evaluate the matching-scale-related systematic uncertainties. The generated samples are reweighted to the MSTW2008LO [23] PDF set.

The MADGRAPH default choice for the renormalisation and factorisation scales is used: the scales are set to the geometric average of $m^2 + p_\text{T}^2$ of the two DM particles, where m denotes their mass. Events with DM masses between 10 and 1300 GeV are simulated for the six different effective operators mentioned above (C1, C5, D1, D5, D9, D11) at LO.

To study the transition between the effective field theory and a simple version of a renormalisable model (a *Simplified Model*) for Dirac fermion DM coupling to Standard Model particles via a new mediator particle Z', events for this Simplified Model are generated in MADGRAPH. For each DM mass, mediator particle masses M_med between 50 GeV and 30 TeV are considered, each for two values of the mediator particle width ($\Gamma = M_\text{med}/3$ and $M_\text{med}/8\pi$).

Pile-Up and Detector Simulation

The effect of pile-up is emulated by overlaying several minimum-bias events, generated with PYTHIA 8, onto the hard scattering. The pile-up distribution is adjusted according to the run conditions and the instantaneous luminosity present during data taking.

In the end, the MC samples are processed using a simulation of the ATLAS detector. The background samples use the *full simulation*, the signal samples rely on the *fast simulation*. The difference between the full and the fast detector simulation is found to be negligible for this analysis.

7.3 Event Selection

7.3.1 Reconstructed Objects

Jets with $p_T > 30$ GeV and $|\eta| < 4.5$ are considered in the analysis. Reconstructed leptons (electrons or muons) are used to reject events that arise from leptonic background processes. Furthermore, leptons in the final state are selected to define control samples. Muon are considered in the lepton veto, if they fulfil $p_T > 7$ GeV and $|\eta| < 2.5$ and if they are isolated: the sum of the transverse momenta of the tracks not associated with the muon in a cone of size $\Delta R = \sqrt{(\Delta\eta)^2 + (\Delta\phi)^2} = 0.2$ around the muon direction is required to be less than 1.8 GeV. The muon p_T requirement is tightened to $p_T > 20$ GeV in the definition of the $W(\to \mu\nu)$+jets and $Z/\gamma^*(\to \mu^+\mu^-)$+jets control regions.

In order to be considered in the lepton veto electrons are required to have $p_T > 7$ GeV and $|\eta| < 2.47$, and to pass the medium electron shower shape and track selection criteria described in Ref. [37]. Possible overlaps between electrons and jets are resolved by discarding the jet if the radial distance ΔR from any electron is less than 0.2. For the definition of the $Z/\gamma^*(\to e^+e^-)$+jets and $W(\to e\nu)$+jets control regions, the electron p_T requirement is increased to $p_T > 20$ GeV and electrons in the transition region between calorimeter sections $1.37 < |\eta| < 1.52$ are excluded. The selection is further tightened in case of the $W(\to e\nu)$+jets control region that is used to estimate the irreducible $Z(\to \nu\bar{\nu})$+jets background. To achieve a cleaner sample, electrons are required to pass tight [37] electron shower shape and track selection criteria, their p_T threshold is raised to 25 GeV, and they need to be isolated: the sum of the transverse momenta of the tracks not associated with the electron in a cone of radius $\Delta R = 0.3$ around the electron direction is required to be less than 5% of the electron p_T. An analogous isolation criterion, based on the calorimeter energy deposits not associated with the electron, is also applied.

A purely calorimeter based E_T^{miss} is used in this analysis. Consequently, muons are not considered in the E_T^{miss} calculation.

7.3.2 Preselection

A common preselection is applied to all events considered in the analysis: a reconstructed primary vertex with at least two associated tracks with $p_T > 0.4$ GeV is required. If more than one vertex is found, the vertex with the largest summed p_T^2 of the associated tracks is chosen. Events are required to fulfil $E_T^{miss} > 150$ GeV to ensure that the E_T^{miss} trigger is fully efficient, and to contain at least one jet with $p_T > 120$ GeV and $|\eta| < 2.0$. Additional jets with $p_T > 30$ GeV and $|\eta| < 4.5$ are allowed to be present in the final state. In order to reduce contributions from top and multijet backgrounds, where large E_T^{miss} mainly results from mis-measurements of jets, the direction of E_T^{miss} and the jets is required to be separated by $\Delta\phi(\text{jet}, \vec{p}_T^{miss}) > 1.0$.

Events containing possibly mis-measured jets are rejected: any jet above $p_T > 20$ GeV and within $|\eta| < 4.5$ needs to be consistent with originating from the collision vertex, constraints on the electromagnetic fraction in the calorimeter, calorimeter sampling fraction, and the so-called charged fraction are imposed [38]. If any of the jets is reconstructed close to a region of the calorimeter that is known to be only partially instrumented, the event is rejected. Additional requirements based on the timing and the pulse shape of the cells in the calorimeter are applied to suppress coherent noise and electronic noise bursts in the calorimeter producing anomalous energy deposits [39]; the conditions have a negligible effect on the signal efficiency. These requirements are tightened on the leading jet in the event to reject possibly remaining contributions from beam-related backgrounds and cosmic rays.

Events containing identified muons or electrons with $p_T > 7$ GeV are vetoed. In addition, events with isolated tracks with $p_T > 10$ GeV and $|\eta| < 2.5$ are vetoed to reduce background from non-identified leptons (e, μ or τ) in the final state.

7.3.3 Veto on Isolated Tracks

The largest reducible backgrounds in the Monojet analysis arise from electroweak processes where the muon or electron is not rejected by the lepton veto or where a hadronically decaying tau lepton is involved. Especially the background coming from $W(\rightarrow \tau\nu)$ amounts to 25% in a signal region-like selection with E_T^{miss} larger than 120 GeV and to 16% above E_T^{miss} of 350 GeV. A large fraction of this background component is coming from events where the τ decays hadronically and hence looks similar to a jet. There is a balance to find between possible background suppression and the additional systematic uncertainty introduced by adding (or tightening) a requirement.

In the following, a veto on isolated tracks, developed specifically for this analysis, is presented. While the electroweak backgrounds are reduced, the signal efficiency is high ($\epsilon > 95\%$), and only small systematic uncertainties are introduced by the veto ($<1\%$). The idea is to remove events with isolated tracks in order to reject events that contain a jet from a hadronic tau decay, since such tau jets typically contain less tracks than normal jets. Furthermore, events containing a leptonic tau decay or leptons that escape the veto can be efficiently reduced.

In this section, the set of cuts applied for the plots corresponds to the preselection, detailed in Sect. 7.3.2, and a lower cut on E_T^{miss}, making the selection signal-region-like. Furthermore, a veto on more than two jets applied, if not specified otherwise.[3]

7.3.3.1 Candidate Variables for Track Isolation

A certain quality of the considered tracks has to be required in order to make an isolated-track veto less dependent on pile-up. The quality cuts considered in the following require a p_T above 10 GeV, $|\eta| < 2.5$ and small impact parameters with respect to the primary vertex ($|z_0| < 2$ mm, $|d_0| < 1$ mm). Furthermore, at least five hits in the tracker and $\chi^2/d.o.f. < 3$ are required to ensure a sufficiently good quality of the track fitting.

The figures of merit considered for the study of different variables, their discrimination power and their performance are pile-up dependence, signal efficiency and background rejection. Three types of variables are studied:

- **p_T cone track**: the sum of the transverse momenta of tracks in a cone around a track of interest,
- **p_T cone calo**: the same quantity, but constructed from calorimeter information,
- **cone n**: the number of tracks around the track of interest.

In addition, the same variables scaled by the p_T of the track of interest are considered. The naming convention for the studied variables is the following: `coneXX_YYg_ZZ`, where XX is the radial size of the cone centred around the particle of interest, YY is the momentum threshold considered for the tracks (in GeV), and ZZ is either `track` to denote p_T *cone track*, `calo` for p_T *cone calo*, and `n` for *cone n*.

7.3.3.2 Signal Efficiency and Background Rejection

The behaviour of the different variables are studied for the irreducible background $Z(\rightarrow \nu\bar{\nu})$ as a proxy for a signal-like topology and then compared to all other electroweak background processes. Figure 7.2 illustrates that tracks in signal-like $Z(\rightarrow \nu\bar{\nu})$ events are much less isolated than for the electroweak background samples. Figure 7.3 shows that any veto on isolated tracks improves the ratio of signal over background, since the efficiency for the background processes drops faster than the one for $Z(\rightarrow \nu\bar{\nu})$ for all studied isolation variables.

---

[3] As will be seen in the following, this jet veto is removed, following the DM optimisation. The performance of the isolated-track veto is not changed significantly when the veto is not applied.

7.3 Event Selection

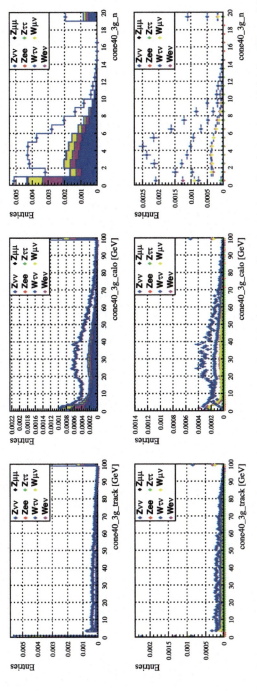

Fig. 7.2 Distribution of track p_T (left) and calorimeter p_T (middle) based track isolation as well as the number of tracks in a cone (right) for different Monte Carlo samples (ALPGEN). As an example, a cone of 0.4 and a minimal p_T of tracks in the cone of 3 GeV is shown here. It can be seen from the stacked (top) and overlaid (bottom) plots that signal-like $Z(\to \nu\bar{\nu})$ is less isolated than other electroweak backgrounds

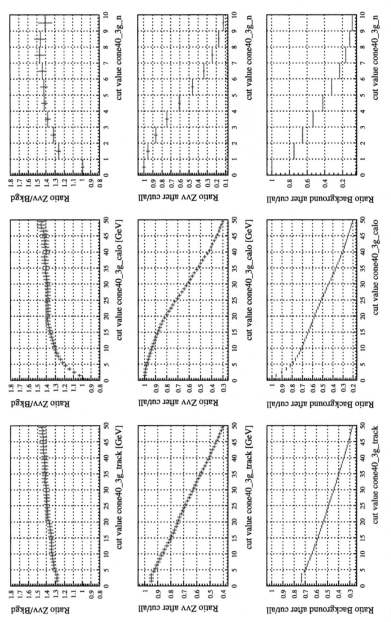

Fig. 7.3 Different cut values for all three variables, cone40_3g_track (left), cone40_3g_calo (middle), cone40_3g_n (right) are investigated. The ratio of the signal-like $Z(\to \nu\bar{\nu})$ over all other backgrounds (top), the efficiency of the cut for signal-like $Z(\to \nu\bar{\nu})$ events (middle) and for other backgrounds (bottom) are studied in Monte Carlo samples

7.3 Event Selection

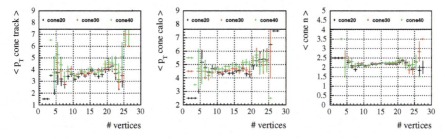

Fig. 7.4 Average track isolation variable in data (period L3, 219 pb^{-1}), as a function of the number of primary vertices

Efficiencies for both $Z(\to \nu\bar{\nu})$ and backgrounds are found to be higher for larger cone sizes and p_T-scaled variables. The ratio of signal(-like $Z(\to \nu\bar{\nu})$) over background, S/B, is higher for smaller cone sizes. If only tracks in the cone above a certain p_T threshold are considered, more $Z(\to \nu\bar{\nu})$ and background events are rejected. Applying such a minimum p_T threshold improves the S/B, especially for the variable counting the number of tracks in the cone.

7.3.3.3 Pile-up Stability

Since track isolation can be affected by pile-up interactions in the event, one of the most crucial requirements for a variable used for a track veto is to be largely independent of the number of primary vertices observed in the collision.[4] While this is true for all of the studied variables, *cone n* performs especially well as can be seen in Fig. 7.4.

Taking also the previous findings into account, the variable `cone40_3g_n`, namely the number of tracks with a p_T of more than 3 GeV in a cone of 0.4 around the track of interest was chosen for this analysis.

7.3.3.4 Choice of Cut Value

The distribution of the quantity `cone40_3g_n` for different backgrounds (Fig. 7.5) shows that the main difference between signal and background is concentrated in the first bin, i.e. the main difference between signal and background is whether or not a track can be found in the vicinity of the track of interest. Therefore, the veto is defined to reject events that contain tracks without another track above 3 GeV in their vicinity: `cone40_3g_n > 0`.

---

[4]The number of vertices is counted from reconstructed vertices with at least one associated track.

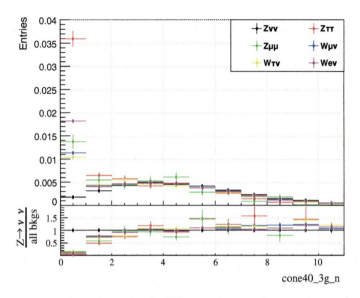

Fig. 7.5 Comparison of `cone40_3g_n` for different backgrounds in Monte Carlo (ALPGEN). The ratio between $Z(\to \nu\bar{\nu})$ and all backgrounds is shown in the bottom panel. The difference between signal-like topologies and background is concentrated in the first bin

7.3.3.5 Performance of the Track Veto

To study the efficiency of the track veto and the improvements it can bring to the analysis, event numbers after all preselection cuts including the lepton vetoes (see Sect. 7.3.2) and above a certain E_T^{miss} are compared before and after the track veto. Figure 7.6 shows the signal over background ratio with and without the track veto applied, showing that an improvement of about 10% is seen over the whole range

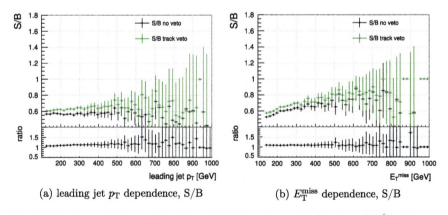

(a) leading jet p_T dependence, S/B (b) E_T^{miss} dependence, S/B

Fig. 7.6 Signal to background ratio with and without track veto, as a function of E_T^{miss} and p_T

7.3 Event Selection

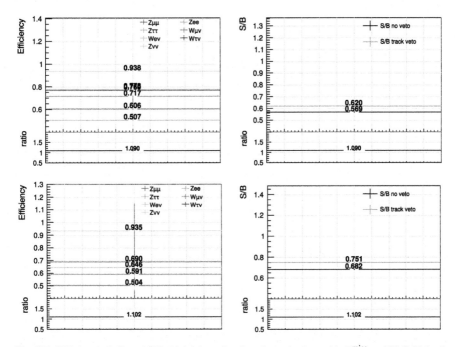

Fig. 7.7 Efficiency (left) and S/B (right) in a signal region selection with $E_T^{miss} > 150$ GeV (top) and with $E_T^{miss} > 350$ GeV (bottom)

of E_T^{miss} and jet p_T. Here and in the following the term signal over background ratio (S/B) refers to the ratio of $Z(\to \nu\bar{\nu})$ events over the sum of events from $Z(\to \nu\bar{\nu})$ and all other electroweak backgrounds. As seen in Fig. 7.7, the efficiency of the track veto for $Z(\to \nu\bar{\nu})$ events with $E_T^{miss} > 150$ GeV lies around 95%, meaning that applying the veto approximately 5% of signal events get rejected. The efficiencies for all other electroweak backgrounds lie below. $Z/\gamma^*(\to \mu^+\mu^-)$, $W(\to \mu\nu)$, $W(\to \tau\nu)$ group around 70–75%, whereas the efficiencies for $Z/\gamma^*(\to \tau^+\tau^-)$ and $W(\to e\nu)$ are even lower. The improvement of S/B is of the order of 7%.

Events with $E_T^{miss} > 350$ GeV (Fig. 7.7) behave similarly to the tested events with lower E_T^{miss}, with efficiencies that are higher for $Z(\to \nu\bar{\nu})$ than for other backgrounds. The $W(\to \tau\nu)$ background rejection is not further improved, which can be understood, given that a high-p_T hadronically-decaying tau looks very similar to a "normal" jet. The improvement of S/B is also similar to the one at lower E_T^{miss}, namely about 9%.

7.3.3.6 Systematic Effects

Since the electroweak backgrounds entering the SRs are estimated in $W(\to \ell\nu)$ and $Z/\gamma^*(\to \ell^+\ell^-)$ control regions (see Sect. 7.1), the veto efficiencies of these

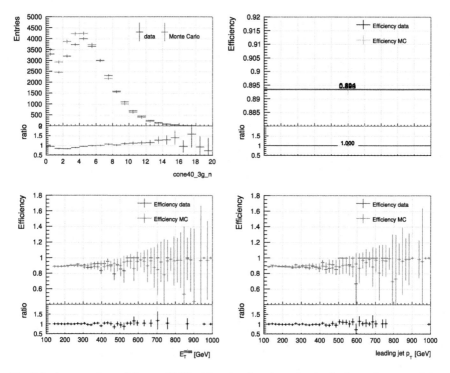

Fig. 7.8 A comparison of data and MC with a signal region selection is shown in terms of the distribution of the isolation variable (upper left) and the total efficiency of the track veto (upper right), as well as the E_T^{miss} (lower left) and jet p_T (lower right) dependence of the track veto efficiencies

processes in MC need to be thoroughly checked and compared to the ones in data to avoid systematic effects on the transfer factors entering the data-driven background estimation. In the following, the results for the muon selections are presented. It was confirmed that the same qualitative and quantitative conclusions hold as well for the electron selections.

Modelling of the Veto Efficiency in the SR Figure 7.8 compares MC and data after applying a SR-like selection. The isolation variable is reasonably well modelled, although the ratio presents a slope with discrepancies of up to 20%. Relevant for the analysis is whether the efficiency of the isolated-track veto is well modelled. There, the ratio of data and MC efficiencies is consistent with unity within statistical uncertainties. Furthermore, the data-MC ratio shows no trend with E_T^{miss} or jet p_T.

Similarity of Signal and CR Events In order to successfully apply the transfer factor method, the MC efficiency in $Z(\to \nu\bar{\nu})$ with a signal region selection should be the same as for the other backgrounds when applying a control region selection and explicitly excluding the CR lepton(s) from the veto. This comparison is shown in

7.3 Event Selection

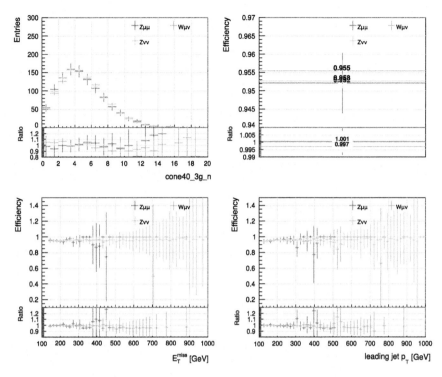

Fig. 7.9 A comparison is shown of $Z(\to \nu\bar{\nu})$, $W(\to \mu\nu)$ and $Z/\gamma^*(\to \mu^+\mu^-)$ events in terms of the distribution of the isolation variable (upper left) and the total veto efficiency (upper right), as well as the E_T^{miss} (lower left) and jet p_T (lower right) dependence of the veto efficiencies. The lepton(s) are excluded from the track veto

Fig. 7.9 and proves that indeed the SR and CR events–apart from the lepton(s)–look very similar in $Z(\to \nu\bar{\nu})$, $Z/\gamma^*(\to \ell^+\ell^-)$ and $W(\to \ell\nu)$ in terms of track isolation.

Modelling of Non-leptonic Part of Control Region Events Figure 7.10 compares MC and data in the $W(\to \mu\nu)$ CR, excluding the muon from the track veto. The track veto efficiency is well modelled although small discrepancies can be noticed. The veto efficiencies differ by 0.2%. No trend with E_T^{miss} or jet p_T is observed. For completeness, the modelling was also checked in the $Z/\gamma^*(\to \mu^+\mu^-)$ CR (Fig. 7.11), which is cleaner than the $W(\to \mu\nu)$ CR. The picture is similar, the veto efficiencies differ by 1.2%.

Leptonic Part of the Event Figure 7.12 compares MC and data in the $W(\to \mu\nu)$ CR. Here, the muon is explicitly included when the veto is applied in order to probe the modelling of the lepton isolation. The isolation is well modelled although a small discrepancy can be noticed. The veto efficiencies differ by 0.6%, meaning a relative difference of 27%. Again, no trend with E_T^{miss} or jet p_T is observed. The modelling was also tested in the $Z/\gamma^*(\to \mu^+\mu^-)$ CR (Fig. 7.12). The findings are in agreement with the ones from the $W(\to \mu\nu)$ CR, the veto efficiencies differ by 2%.

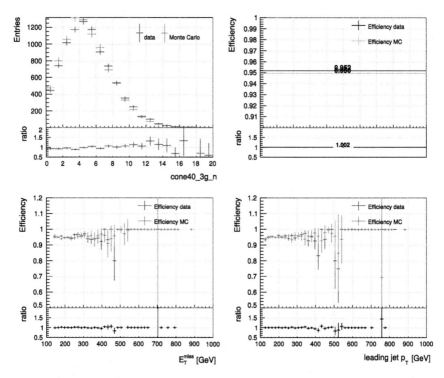

Fig. 7.10 The plots show the comparison of data and MC in the $W(\to \mu\nu)$ CR in terms of the distribution of the isolation variable (upper left) and the total veto efficiency (upper right), as well as the E_T^{miss} (lower left) and jet p_T (lower right) dependence of the veto efficiencies. The lepton was excluded from the track veto

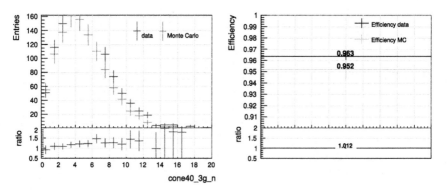

Fig. 7.11 The plots show the comparison of data and MC in the $Z/\gamma^*(\to \mu^+\mu^-)$ CR in terms of the distribution of the isolation variable (left) and the total veto efficiency (right). The lepton was excluded from the track veto

7.3 Event Selection

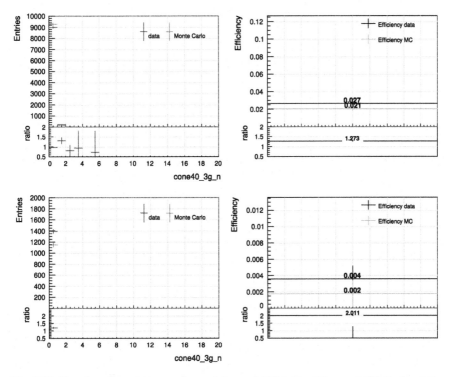

Fig. 7.12 The plots show the comparison of data and MC in the $W(\to \mu\nu)$ CR (top) and the $Z/\gamma^*(\to \mu^+\mu^-)$ CR (bottom) in terms of the distribution of the isolation variable (left) and the total veto efficiency (right)

Tau leptons and Multijets Due to the limited statistics in the available multijet MC samples, the remaining QCD background in the signal regions is estimated from data (see Sect. 7.4.2). The track veto efficiency is measured in multijet-enriched data events and found to be 0.903 for the jet-veto selection and 0.871 for the inclusive selection. These factors are applied on the final QCD estimate done without applying the track veto. In order to probe the modelling of tau leptons in particular a new control region was defined by lowering the E_T^{miss} and restricting it to $100\,\text{GeV} < E_T^{miss} < 150\,\text{GeV}$. Reducing it further is not possible, since effects of the trigger turn-on would enter. The upper cut avoids having overlap with the signal region definition. The results shown in Fig. 7.13 confirm a reasonable modelling of the relevant quantities also in this region.

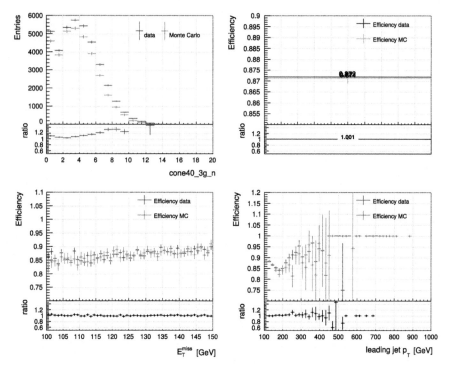

Fig. 7.13 The plots show the comparison of data and MC in the constructed tau CR in terms of the distribution of the isolation variable (upper left) and the total veto efficiency (upper right), as well as the E_T^{miss} (lower left) and jet p_T (lower right) dependence of the veto efficiencies

7.3.4 Cut Optimisation for Dark Matter Signals

In order to optimise the sensitivity to DM signals, several modifications of the monojet-like signal selection were tested. The focus is put on both the D5 (vector) and the D11 (scalar, gluon) operator, since they lead to rather different E_T^{miss} distributions. DM masses of 100, 400 and 1000 GeV are considered.

As a figure of merit for the improvement in sensitivity the following expression, taken from Ref. [40], is used:

$$\frac{\varepsilon(t)}{S_{min}}. \qquad (7.1)$$

$\varepsilon(t)$ denotes the signal efficiency corresponding to a specific set of cuts t. The denominator, S_{min} is given by:

$$S_{min} = \frac{a^2}{8} + \frac{9b^2}{13} + a\sqrt{B} + \frac{b}{2}\sqrt{b^2 + 4a\sqrt{B} + 4B}. \qquad (7.2)$$

7.3 Event Selection

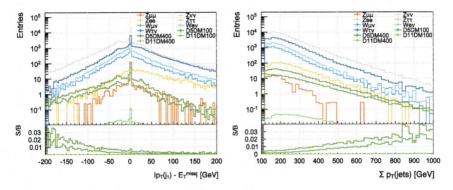

Fig. 7.14 The comparison of different signal and background distributions. The difference between leading jet p_T and E_T^{miss} is shown on the left, the sum of jet p_T's on the right. The distributions reveal that signal events tend to have a larger jet p_T sum and a larger difference between leading jet p_T and E_T^{miss}

It relates the number of background events passing the selection, $B(t)$, and the number of standard deviations required for discovery, $a = 5$, and exclusion, $b = 2$, where the latter corresponds to a limit set at 95% confidence level. In this way, the selection is optimised for both discovery and exclusion. Samples of $Z(\to \nu\bar{\nu})$+jets, $W(\to \ell\nu)$+jets, top and diboson processes are considered as backgrounds, while contributions from multi-jet and $Z/\gamma^*(\to \ell^+\ell^-)$+jets are neglected. A preselection is applied, leaving the number of jets unrestricted and not yet cutting on the $\Delta\phi$ between the jet(s) and E_T^{miss}. The leading jet p_T and E_T^{miss} are required to be above 120 GeV, and events containing electrons or muons are vetoed.

As can be seen in Fig. 7.14, the signal is more asymmetric in E_T^{miss} and jet multiplicity, while in the $Z(\to \nu\bar{\nu})$ background the E_T^{miss} is balanced by the leading jet. Consequently, not limiting the number of jets in the event can increase the signal acceptance and hence the sensitivity. This is supported by the fact that signal events tend to have a larger sum p_T of jets, as illustrated in Fig. 7.14, indicating that the energy scale of signal events tends to be higher.

However, the jet veto reduces backgrounds that do not rely on initial-state radiation jets and hence can easily reach large jet multiplicities, such as multi-jet or top processes. To attenuate the increase of such backgrounds, the cut on $\Delta\phi$ between the sub-leading jet and E_T^{miss}, which was already used in previous rounds of the analysis, is extended to all jets such that the minimum $\Delta\phi$ between the E_T^{miss} and any jet is required to be larger than a certain value (0.5 by default). The improvement from releasing the jet veto can be seen from the numbers in Table 7.2, where ε/S_{min} is listed for a number of signal points and different cuts on the jet multiplicity. The cut on minimum $\Delta\phi(\text{jets}, E_T^{\text{miss}}) > 0.5$ is included. The numbers show the same trend for all operators and mass points: loosening the jet veto increases the sensitivity.

If more jets are allowed in the final state, the ratio of the leading jet p_T to E_T^{miss} is expected to be smaller in signal events, since there, as discussed above, leading jet p_T and E_T^{miss} are more asymmetric than for the backgrounds. Asymmetric cuts

Table 7.2 Significance measure ε/S_{min} for different jet multiplicities and various signal points. For all operators and masses the same trend is observed: the sensitivity is improved by allowing for larger jet multiplicities.

Operator	m_χ [GeV]	3rd jet veto	4th jet veto	No jet veto
D11	100	193.9	255.1	300.4
	400	217.8	298.8	364.8
D5	100	148.2	174.2	181.9
	400	168.7	202.0	212.0
C1	100	82.5	95.6	98.0
	400	149.4	184.0	197.3
C5	100	152.9	196.8	224.1
	400	202.8	276.1	327.6

for leading jet p_T and E_T^{miss} are hence examined. Figure 7.15 shows the sensitivity measure as a function of the cut values for leading jet p_T and E_T^{miss}. For a given value of E_T^{miss}, an increase in the leading jet p_T cut does not improve the sensitivity. For a chosen value of a cut on leading jet p_T, the optimal value of the E_T^{miss} cut depends strongly on the operator and to a certain extent on the DM mass. In the end, the cut on the leading jet p_T is set to 120 GeV and the inclusive E_T^{miss} cut is increased in steps of 50 GeV, starting at 150 GeV.

7.3.5 Signal Region Definition

Taking the findings on the veto on isolated tracks and from the DM signal optimisation, the preselection described in Sect. 7.3.2 is complemented by a cut on the number of isolated tracks (n_{tracks} (`cone40_3g_n = 0`) = 0) and by a requirement on the ratio of E_T^{miss} and leading jet p_T ($p_{T,jet_1}/E_T^{miss} > 0.5$). The different signal regions are then defined via increasing, inclusive E_T^{miss} thresholds from 150 to 700 GeV. An overview of the selection is given in Table 7.3.

7.4 Background Estimation

The main backgrounds entering the signal selection come from electroweak processes. $Z(\to \nu\bar{\nu})$ as being irreducible presents the largest fraction. $W(\to \ell\nu)$ events can pass the selection in case the lepton rests unidentified or is outside of the acceptance. The MC prediction for these backgrounds is normalised to data in dedicated, background-enriched control regions. The obtained normalisation is then transferred to the signal region. Smaller backgrounds such as $Z/\gamma^*(\to \ell^+\ell^-)$, top and diboson contributions are estimated from MC. The $t\bar{t}$ background prediction is validated using

7.4 Background Estimation

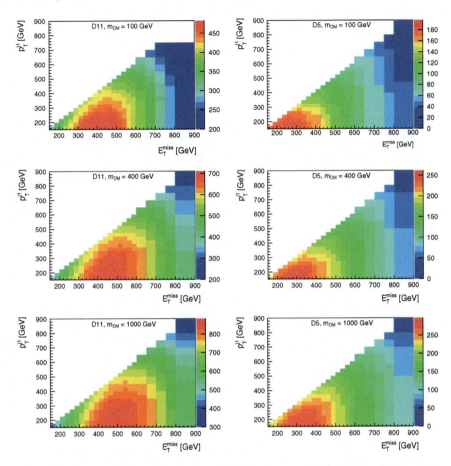

Fig. 7.15 Sensitivity ($\epsilon(t)/S_{min}$) as a function of leading jet p_T and E_T^{miss} for D5 and D11 at three different DM mass points

Table 7.3 Event selection criteria for the monojet-like signal regions

Selection criteria										
Pre-selection	Primary vertex									
	$E_T^{miss} > 150$ GeV									
	Jet quality requirements									
	At least one jet with $p_T > 30$ GeV and $|\eta| < 4.5$									
	Lepton and isolated track vetoes									
Monojet selection	Leading jet $p_T > 120$ GeV and $|\eta| < 2.0$									
	Leading jet $p_T/E_T^{miss} > 0.5$									
	$\Delta\phi(\text{jet}, \vec{p}_T^{\,miss}) > 1.0$									
Signal regions		SR1	SR2	SR3	SR4	SR5	SR6	SR7	SR8	SR9
	E_T^{miss} (GeV)	>150	>200	>250	>300	>350	>400	>500	>600	>700

Table 7.4 Summary of the techniques used to predict the relevant backgrounds in the SRs.

Background process	Method	Control sample
$Z(\to \nu\bar{\nu})$+jets	MC and control samples in data	$Z/\gamma^*(\to \ell^+\ell^-)$, $W(\to \ell\nu)$ ($\ell = e, \mu$)
$W(\to e\nu)$+jets	MC and control samples in data	$W(\to e\nu)$ (loose)
$W(\to \tau\nu)$+jets	MC and control samples in data	$W(\to e\nu)$ (loose)
$W(\to \mu\nu)$+jets	MC and control samples in data	$W(\to \mu\nu)$
$Z/\gamma^*(\to \ell^+\ell^-)$+jets ($\ell = e, \mu, \tau$)	MC only	
$t\bar{t}$, single top	MC only	
Diboson	MC only	
Multijets	Data-driven	
Non-collision	Data-driven	

a top-enriched selection. The estimate of the multijet background relies purely on data, as well as the one of the non-collision background. The background estimation strategy is summarised in Table 7.4.

7.4.1 W/Z+jets Background

In order to constrain the background contribution from electroweak processes, dedicated control regions are defined. They are orthogonal to the signal selection by explicitly requiring identified leptons. Otherwise, the selection requirements are kept identical. This allows to significantly reduce the impact of theoretical and experimental systematic uncertainties. The data in the control regions and MC–based correction factors are used to estimate the electroweak background contributions from W+jets and $Z(\to \nu\bar{\nu})$+jets processes in each of the nine signal regions.

The $W(\to \mu\nu)$+jets background in a specific signal region, $N_{\text{SR}}^{W(\to \mu\nu)}$ is estimated from the $W(\to \mu\nu)$+jets control region via:

$$N_{\text{SR}}^{W(\to \mu\nu)} = \frac{(N_{W(\to \mu\nu),\text{CR}}^{\text{data}} - N_{W(\to \mu\nu),\text{CR}}^{\text{non}-W/Z})}{N_{W(\to \mu\nu),\text{CR}}^{\text{MC}}} \times N_{\text{SR}}^{\text{MC}(W(\to \mu\nu))} \times \xi_\ell \times \xi_{\text{trg}} \times \xi_\ell^{\text{veto}}. \tag{7.3}$$

Here, $N_{\text{SR}}^{\text{MC}(W(\to \mu\nu))}$ denotes the $W(\to \mu\nu)$+jets contribution to the signal region predicted by simulation, $N_{W(\to \mu\nu),\text{CR}}^{\text{data}}$, $N_{W(\to \mu\nu),\text{CR}}^{\text{MC}}$ is the number of events in the $W(\to \mu\nu)$ control region in data and MC, respectively. The expression $N_{W(\to \mu\nu),\text{CR}}^{\text{non}-W/Z}$ includes the contribution from non-W/Z backgrounds entering the control region such

7.4 Background Estimation

as top-quark and diboson processes (estimated from MC), and multijets (data-driven estimate). The correction factors ξ_ℓ, ξ_ℓ^{veto}, and ξ_{trg} account for possible differences in the lepton identification, lepton veto, and trigger efficiencies between data and MC. These differences are generally less than 1%.

Analogously, the $Z(\to \nu\bar{\nu})$ estimate from the $W(\to \mu\nu)$ control region reads:

$$N_{\text{SR}}^{Z(\to \nu\bar{\nu})} = \frac{(N_{W(\to \mu\nu),\text{CR}}^{\text{data}} - N_{W(\to \mu\nu),\text{CR}}^{\text{non}-W/Z})}{N_{W(\to \mu\nu),\text{CR}}^{\text{MC}}} \times N_{\text{SR}}^{\text{MC}(Z(\to \nu\bar{\nu}))} \times \xi_\ell \times \xi_{\text{trg}}. \quad (7.4)$$

Similar expressions are derived also for the other estimations. In addition, the shape of the signal region distributions of E_T^{miss} and leading jet p_T are corrected by bin-by-bin correction factors which are derived analogously for plotting the relevant distributions in the signal regions.

The first term appearing on the right-hand side of the above equations can be seen as a *normalisation factor*, accounting for the difference in counts between data and MC in the CR. The derived normalisation factors vary between about 0.9 and 0.6 for the different processes and as the E_T^{miss} requirement is increased from 150 to 700 GeV. They account for a mis-modelling of the p_T of the W and Z bosons which led to the E_T^{miss} and leading jet p_T distributions being softer in data than in MC (around 30%, above $E_T^{\text{miss}} = 600$ GeV).

For each of the signal regions, four separate sets of such *transfer factors* are considered to constrain the dominant $Z(\to \nu\bar{\nu})$+jets background contribution, namely from $Z/\gamma^*(\to \ell^+\ell^-)$+jets and $W(\to \ell\nu)$+jets control regions. The different $Z(\to \nu\bar{\nu})$+jets background estimates in each signal region are found to be consistent within uncertainties and are statistically combined using the BLUE (Best Linear Unbiased Estimate) [41] method, which takes into account correlations of systematic uncertainties.

Muon Control Regions The E_T^{miss} trigger is based on calorimeter information, energy deposits in the muon system are not included in the algorithm. This allows to keep the trigger strategy unchanged from the SRs for these CRs.

The $W(\to \mu\nu)$ control sample is defined by requiring a muon with $p_T > 20$ GeV. The so-called transverse mass m_T is determined by the lepton (ℓ) and neutrino (ν) p_T and direction as $m_T = \sqrt{2 p_T^\ell p_T^\nu (1 - \cos(\phi^\ell - \phi^\nu))}$, where the (x, y) components of the neutrino momentum are taken to be the same as the corresponding $\vec{p}_T^{\text{miss}}$ components. It is a proxy for the transverse mass of the W and is required to be within 40 GeV $< m_T <$ 100 GeV to enhance the fraction of W events. This region is used to estimate the $Z(\to \nu\bar{\nu})$ and the $W(\to \mu\nu)$ background entering in the SRs. Two muons with $p_T > 20$ GeV and an invariant mass close to the Z boson mass, in the range 66 GeV $< m_{\mu\mu} <$ 116 GeV, are required for the $Z/\gamma^*(\to \mu^+\mu^-)$ control region. This region is used to estimate the $Z(\to \nu\bar{\nu})$ contribution to the signal region backgrounds.

Since the E_T^{miss} flavour used in this analysis only considers calorimeter information and hence does not take into account the muon contribution, the E_T^{miss} in

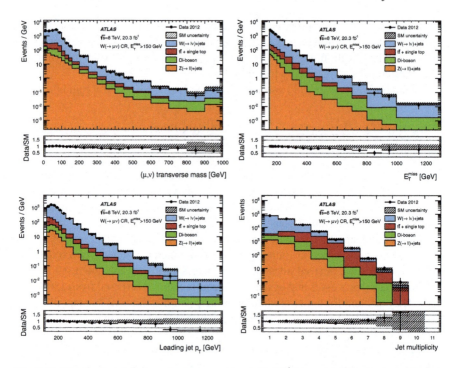

Fig. 7.16 Distribution of the transverse mass m_T (top left), E_T^{miss} (top right), leading jet p_T (bottom left) and jet multiplicity (bottom right) in the $W(\to \mu\nu)$+jets control region for the inclusive SR1 selection, compared to the background expectations. The global normalisation factors are already applied. The error bands in the ratios of data to background expectation include statistical and experimental uncertainties

signal and control region correspond to each other. The most relevant kinematic distributions are shown in Fig. 7.16 for the $W(\to \mu\nu)$ region and in Fig. 7.17 for the $Z/\gamma^*(\to \mu^+\mu^-)$ region. The MC processes are globally normalised to the data, such that the shapes of the different distributions in data and MC can be compared. They prove a reasonable modelling of these key variables in MC, although the E_T^{miss} and leading jet p_T distributions are found to be softer in data, as discussed above.

Electron Control Regions The electron control regions cannot rely on the E_T^{miss} trigger and hence rely on single-electron triggers as described in Sect. 7.2.

The $W(\to e\nu)$+jets control sample that is used to estimate the $Z(\to \nu\bar{\nu})$ background requires one single electron with $p_T > 25$ GeV, a transverse mass in the range 40 GeV $< m_T <$ 100 GeV, and $E_T^{miss} > 25$ GeV. The requirements are optimised to suppress events in which jets are misidentified as electrons. To estimate the $Z(\to \nu\bar{\nu})$ background, the contributions to the E_T^{miss} calculation that are associated to the electron, coming from energy clusters in the calorimeters, are removed.

A looser selection that allows for more statistics is defined for a $W(\to e\nu)$+jets control sample that is used to estimate the $W(\to e\nu)$ and $W(\to \tau\nu)$ backgrounds

7.4 Background Estimation

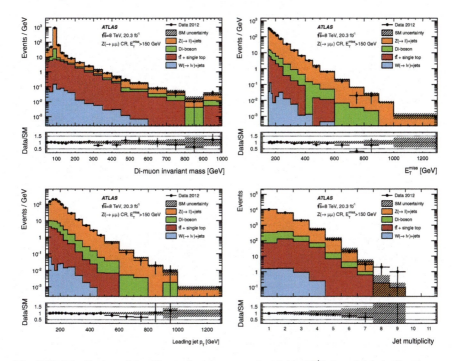

Fig. 7.17 Distribution of the dilepton invariant mass (top left), E_T^{miss} (top right), leading jet p_T (bottom left) and jet multiplicity (bottom right) in the $Z/\gamma^*(\to \mu^+\mu^-)$+jets control region for the inclusive SR1 selection, compared to the background expectations. The global normalisation factors are already applied. The error bands in the ratios of data to background expectation include statistical and experimental uncertainties

entering the signal regions. It is collected with the E_T^{miss}-based trigger, the electron p_T is reduced to $p_T > 20$ GeV and no further cuts on electron isolation and m_T are applied. The E_T^{miss} calculation is not corrected for the contribution from the electron or tau leptons in the final state, as they would contribute to the calorimeter-based E_T^{miss} calculation in the signal regions as well.

In the $Z/\gamma^*(\to e^+e^-)$+jets control region, events are required to have exactly two electrons with $p_T > 20$ GeV and with a dilepton invariant mass around the Z boson mass: $66\,\text{GeV} < m_{ee} < 116\,\text{GeV}$. This region is used to estimate the $Z(\to \nu\bar{\nu})$ contribution to the signal region background. Hence, the E_T^{miss} is corrected for the contributions from the electrons in the calorimeters.

Selected kinematic distributions are shown in Fig. 7.18 for the $W(\to e\nu)$ region and in Fig. 7.19 for the $Z/\gamma^*(\to e^+e^-)$ region. As for the muon CRs a reasonable agreement between data and MC is observed after applying the global normalisation, apart from the E_T^{miss} and leading jet p_T distributions which are found to be softer in data, as discussed above.

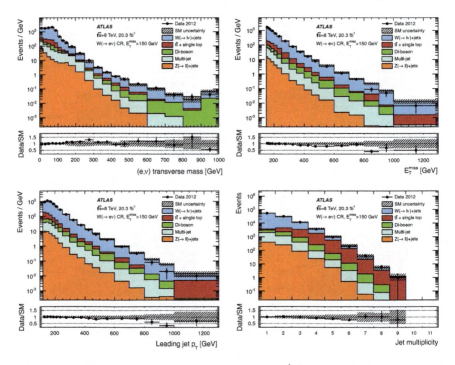

Fig. 7.18 Distributions of the transverse mass m_T (top left), E_T^{miss} (top right), leading jet p_T (bottom left) and jet multiplicity (bottom right) in the $W(\to e\nu)$+jets control region for the inclusive SR1 selection, compared to the background expectations. The global normalisation factors are already applied. The error bands in the ratios of data to background expectation include statistical and experimental uncertainties

7.4.2 Multijet Background

Multijet events from QCD processes can pass the signal region selection with the tight E_T^{miss} requirement only if one or several jets are reconstructed with a wrong energy. It is assumed that such effects are dominated by fluctuations in the detector response and hence the multijet background is estimated using a technique that takes jets measured in data as input and smears their energy and momentum according to the estimated resolution, following the prescription outlined in Ref. [42].

In SR1 and SR2, the multijet background is estimated to be around 2 and 0.7% of the total background, respectively. It is found to be negligible (below 0.5%) for the signal regions with higher E_T^{miss} thresholds.

7.4 Background Estimation

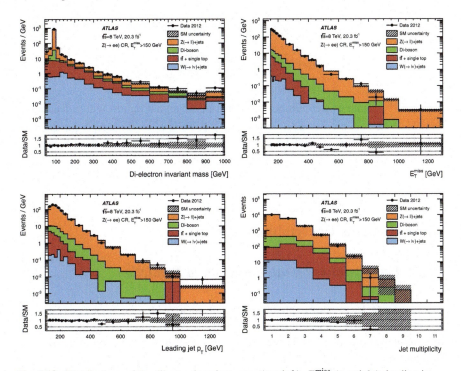

Fig. 7.19 Distributions of the dilepton invariant mass (top left), $E_\mathrm{T}^\mathrm{miss}$ (top right), leading jet p_T (bottom left) and jet multiplicity (bottom right) in the $Z/\gamma^*(\to e^+e^-)$+jets control region for the inclusive SR1 selection, compared to the background expectations. The global normalisation factors are already applied. The error bands in the ratios of data to background expectation include statistical and experimental uncertainties

7.4.3 Non-collision Background

Non-collision events induced by cosmic muons or beam-halo events can look similar to monojet-like event topologies. While the standard selection cuts presented above succeed in reducing the contribution from these backgrounds typically below 1%, their impact needs to be carefully evaluated.

The number of cosmic muon events that pass the selection is estimated in dedicated cosmic datasets and is found to be negligible. Fake jets originating from beam-halo interactions can be tagged as such using information from the spatial alignment of calorimeter and muon system signals [38]. However, this technique has a limited efficiency. Another way to identify beam-halo events is by regarding the timing information of the (fake) jets. Jets produced in the collisions are in time with the bunch crossing while fake jets having their origin in beam-halo interactions are not produced within the collision time window. Jets occurring earlier than 5 ns before the bunch crossing are assumed to be beam-halo fakes. Both approaches are combined to estimate the non-collision background present in the signal regions. It is assessed via:

$$N_{\text{NCB}}^{\text{SR}} = N_{t<-5}^{\text{SR}} \times \frac{N^{\text{tag}}}{N_{t<-5}^{\text{tag}}}, \quad (7.5)$$

where $N_{t<-5}^{\text{SR}}$ denotes the number of events in the signal region with a leading jet outside of the bunch-crossing time window -10 ns $< t < -5$ ns, $N_{t<-5}^{\text{tag}}$ is the number of tagged beam-induced background events with an out-of-time jet and N^{tag} represents all identified events in the signal region.

The non-collision background is confirmed to be negligible in all signal regions.

7.5 Systematic Uncertainties

7.5.1 Uncertainties on the Background Prediction

The transfer factor method explained above allows to reduce the impact of systematic uncertainties on the final estimate significantly. Nevertheless, the results are affected by several experimental and theoretical sources of systematic uncertainties. They are summarised in Table 7.5.

The impact of the modelling of W/Z processes is estimated by varying the renormalisation, factorisation, and parton-shower matching scales that are used within the simulation, as well as by evaluating different PDF sets. Furthermore, NLO electroweak corrections, as described in [43–45], are considered. In order to determine the uncertainty on the MC diboson yields, generator and parton shower differences are taken into account, as well as variations of the renormalisation and factorisation scales. Furthermore, different sets of PDFs are tested and parameters for initial- and final state radiation are varied. The uncertainty on the MC estimate of top processes is evaluated using a dedicated $t\bar{t}$ validation region whose selection is close to the signal regions but enhanced in top processes. It requires $\Delta\phi(\vec{p}_{\text{T}}^{\text{miss}}, \text{jet}) > 0.5$ and two b-tagged jets within $|\eta| < 2.4$. The observed difference between data and simulation amounts to 20% for SR1 and up to 100% of the $t\bar{t}$ yield for SR7 and SR9. The uncertainty on the amount of multijet background entering the signal regions is conservatively taken to be 100% of the multijet yield. In addition, multijet and γ+jets events that may enter the $W(\to e\nu)$+jets control region, possibly affect the $Z(\to \nu\bar{\nu})$ background estimate. This is found to be only relevant for the highest-$E_{\text{T}}^{\text{miss}}$ signal region.

Combining these effects and also considering the statistical uncertainties in the control regions the total background expectation is found to be determined to a precision ranging from 2.7% for SR1 and 6.2% for SR7 to 14% for SR9.

7.5 Systematic Uncertainties

Table 7.5 Systematic uncertainties on signal and background yields

Background			
Experimental		Jet energy scale and resolution	0.2–3%
		E_T^{miss} reconstruction	0.2–1%
		Lepton properties	1.4–2%
		Trigger efficiency	0.1% (SR1)
Theoretical		W/Z modelling	1–3%
		Top modelling	0.7–4%
		Diboson modelling	0.7–3%
Other		Multijet estimate	2% (SR1), 0.7% (SR2)
		Multijet and γ+jets in $W(\to e\nu)$ CR	1% (SR9)
Signal			
Acceptance × efficiency		Jet energy scale and resolution, E_T^{miss} reconstruction	1–10%
		Beam energy	3%
		Luminosity	2.8%
		PDF choice	5–29%
		Renormalisation/factorisation scales	3%
		Parton matching scale	5%
Cross section		Beam energy	2–9%
		Renormalisation/factorisation scales	2–17% (D1, D5, D9), 40–46% (C5, D11)
		PDF choice	19–70% (D1, D11, C5), 5–36% (D5, D9) increasing with DM mass

7.5.2 Signal Systematic Uncertainties

For an accurate interpretation of the analysis results it is also important to evaluate the systematic uncertainties that affect the simulated signal scenarios. As for the backgrounds, the uncertainties are computed separately for each signal region. The estimate includes experimental and theoretical sources, where the latter are evaluated by varying several parameters of the simulation.

Uncertainties on jet energy scale and resolution as well as on the E_T^{miss} reconstruction are considered as sources of experimental systematics alongside with the negligible (below 1%) contributions related to the jet quality and the track veto selection. Furthermore, the uncertainties on the beam energy, the integrated luminosity (2.8%) and the trigger efficiency (1% in SR1) are taken into account.

Theoretical uncertainties on the simulation details are estimated by varying the initial- and final-state radiation parameters of the parton showering, i.e. via comparing simulated samples with enhanced and suppressed parton emission (using $\alpha_s(2p_T)$

Table 7.6 Observed event yield in data and background expectations in the signal regions. The errors on the background expectations include statistical and systematic uncertainties. Since the uncertainties for the different background processes can generally be correlated, the total uncertainty on the background prediction is not necessarily given by the sum of their squares

Signal Region	SR1	SR2	SR3	SR4	SR5	SR6	SR7	SR8	SR9
Observed events	364378	123228	44715	18020	7988	3813	1028	318	126
SM expectation	372100 ± 9900	126000 ± 2900	45300 ± 1100	18000 ± 500	8300 ± 300	4000 ± 160	1030 ± 60	310 ± 30	97 ± 14
$Z(\to \nu\bar{\nu})$	217800 ± 3900	80000 ± 1700	30000 ± 800	12800 ± 410	6000 ± 240	3000 ± 150	740 ± 60	240 ± 30	71 ± 13
$W(\to \tau\nu)$	79300 ± 3300	24000 ± 1200	7700 ± 500	2800 ± 200	1200 ± 110	540 ± 60	130 ± 20	34 ± 8	11 ± 3
$W(\to e\nu)$	23500 ± 1700	7100 ± 560	2400 ± 200	880 ± 80	370 ± 40	170 ± 20	43 ± 7	9 ± 3	3 ± 1
$W(\to \mu\nu)$	28300 ± 1600	8200 ± 500	2500 ± 200	850 ± 80	330 ± 40	140 ± 20	35 ± 6	10 ± 2	2 ± 1
$Z/\gamma^*(\to \mu^+\mu^-)$	530 ± 220	97 ± 42	19 ± 8	7 ± 3	4 ± 2	3 ± 1	2 ± 1	1 ± 1	1 ± 1
$Z/\gamma^*(\to \tau^+\tau^-)$	780 ± 320	190 ± 80	45 ± 19	14 ± 6	5 ± 2	2 ± 1	0 ± 0	0 ± 0	0 ± 0
$t\bar{t}$, single top	6900 ± 1400	2300 ± 500	700 ± 160	200 ± 70	80 ± 40	30 ± 20	7 ± 7	1 ± 1	0 ± 0
Dibosons	8000 ± 1700	3500 ± 800	1500 ± 400	690 ± 200	350 ± 120	183 ± 70	65 ± 35	23 ± 16	8 ± 7
Multijets	6500 ± 6500	800 ± 800	200 ± 200	44 ± 44	15 ± 15	6 ± 6	1 ± 1	0 ± 0	0 ± 0

7.5 Systematic Uncertainties 127

and $\alpha_s(p_T/2)$, respectively) to the nominal ones. Further, the impact of the choice of a specific PDF set alongside with the value of the strong coupling inserted in the simulation, $\alpha_s(m_Z)$, is evaluated: the envelope of CT10, MRST2008LO and NNPDF21LO is taken as uncertainty. In addition, the applied values for the renormalisation and the factorisation scales are varied by factors of one half and two in MADGRAPH. The matching scale between matrix-element calculation and parton-shower modelling is varied in the same way. The choice of PDF set and renormalisation and factorisation scales as well as the uncertainty on the beam energy also affects the predicted signal cross section, which is taken into account. The uncertainties related to the hadronic showering (α_s value and initial- and final-state radiation), the trigger efficiency and jet quality are smaller than 1% and hence considered negligible.

A summary of the systematic uncertainties on the signal samples is given in Table 7.5.

7.6 Results

The event yield in data is found to agree with the background prediction in all signal region. The largest deviation observed amounts to 1.7σ and is seen in the signal region requiring the highest E_T^{miss} (SR9). The results are summarised in Table 7.6, distributions of selected variables are compared between data and background expectation in Figs. 7.20 and 7.21. While the most inclusive SR (SR1) sees over 360000 events, the tightest SR (SR9) observes 126 events.

7.7 Interpretation

7.7.1 Model-Independent Limits

The measured number of events in data and the predicted background in the signal regions is used to calculate upper limits on the visible cross section, given by the product of the signal cross section, its acceptance and its efficiency, $\sigma \times A \times \epsilon$. This is done following the CL_s modified frequentist approach [46]. While a confidence level of 95% is common for collider searches, direct Dark Matter detection experiments often quote limits at 90% confidence level. Hence, both cases are considered. Their calculation includes the systematic uncertainties described above.

The signal-like selection efficiencies estimated via a $Z(\to \nu\bar{\nu})$ sample range from 88% for SR1 and 83% for SR3 to 82% for SR7 and 81% for SR9.[5] Visible cross sections above 599 fb (726 fb) are excluded by SR1 at 90% CL (95% CL). SR9 limits

---

[5]The selection efficiencies are quoted for the irreducible background instead of one (or several) specific signal models. Since signals in general are very similar to $Z(\to \nu\bar{\nu})$ these numbers can be widely applied to many signal models. Furthermore, it is straight-forward to compare a specific

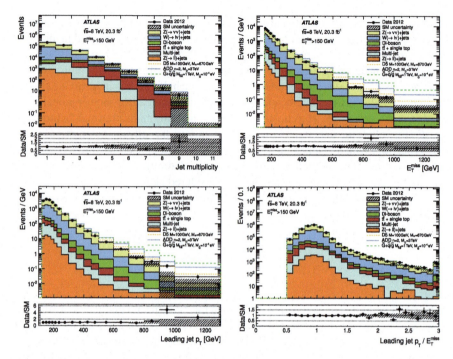

Fig. 7.20 Distributions in data compared to the background expectation of the jet multiplicity (top left), E_T^{miss} (top right), leading jet p_T (bottom left), and the ratio of leading jet p_T and E_T^{miss} (bottom right) in SR1. The normalisation obtained from the $W(\to \mu\nu)$ control region is applied to the $Z(\to \nu\bar{\nu})$ background. The dashed error bands include both the statistical and systematic uncertainties on the background prediction

$\sigma \times A \times \epsilon$ to be smaller than 2.9 fb (3.4 fb) at 90% CL (95% CL). The observed and expected model-independent limits for all signal regions are summarised in Table 7.7.

7.7.2 Dark Matter Pair Production

By considering the visible cross section of signals from DM pair production determined from simulated samples, the model-independent limits can be translated into limits on the signal model parameters.

---

signal model to $Z(\to \nu\bar{\nu})$ and hence determine the relevant signal efficiencies from the $Z(\to \nu\bar{\nu})$ values, if necessary.

7.7 Interpretation

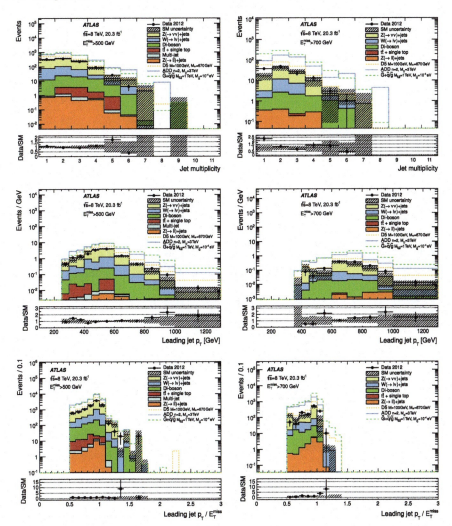

Fig. 7.21 Distributions in data compared to the background expectation of (from top to bottom) the jet multiplicity, E_T^{miss}, leading jet p_T, and the ratio of leading jet p_T and E_T^{miss} in SR7 (left) and SR9 (right). The normalisation obtained from the $W(\to \mu\nu)$ control region is applied to the $Z(\to \nu\bar{\nu})$ background. The dashed error bands include both the statistical and systematic uncertainties on the background prediction

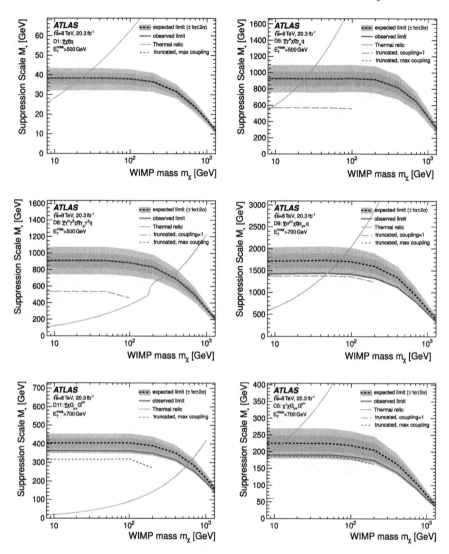

Fig. 7.22 95% CL lower limits on the suppression scale Λ (here denoted as M_*) are shown as a function of the DM mass m_χ for different interaction operators: (from left to right and top to bottom) D1, D5, D8, D9, D11 and C5. In each case, the most sensitive signal region (SR7 for D1, D5, D8; SR9 for D9, D11 and C5) is considered. The expected and observed limits are shown as dashed black and solid blue lines, respectively. The rising green lines denote the values of cut-off scales Λ at which DM particles of the given mass would lead to the current relic density as measured by WMAP [47], assuming that DM in the early universe exclusively annihilated via the considered operator. The purple long-dashed line shows the modified limit on the cut-off scale when imposing a validity constraint obtained assuming coupling strengths of one, the red dashed thin lines are those for maximum coupling strength (see text for further details)

7.7 Interpretation

Table 7.7 Model-independent limits on $\sigma \times A \times \epsilon$ at 90% CL and 95% CL.

Signal region	Upper limits on $\sigma \times A \times \epsilon$ [fb]			
	90% CL		95% CL	
	Observed	Expected	Observed	Expected
SR1	599	788	726	935
SR2	158	229	194	271
SR3	74	89	90	106
SR4	38	43	45	51
SR5	17	24	21	29
SR6	10	14	12	17
SR7	6.0	6.0	7.2	7.2
SR8	3.2	3.0	3.8	3.6
SR9	2.9	1.5	3.4	1.8

7.7.2.1 Effective Field Theory Interpretation

Within the effective field theoretic approach the limit is put on the EFT cut-off scale Λ at 95% CL for different interaction operators in dependence on the DM mass. For each interaction operator, the signal region yielding the best expected limits is considered, namely SR4 for C1, SR7 for D1, D5, D8, and SR9 for C5, D9, D11. The resulting limit curves for these effective operators are shown in Fig. 7.22.

The experimental uncertainties on the jet energy scale and resolution, as well as the one on the E_T^{miss} reconstruction enter the limit calculation as a single, fully correlated uncertainty on signal and background event yields, while beam energy and luminosity uncertainty are only considered for the signal expectation.

Generally, experimental and theoretical systematic uncertainties that alter the shape of the E_T^{miss} distribution and therefore the estimated signal acceptance are considered in the calculation of the limits on Λ where those only affecting the overall normalisation by altering the cross section determine the green and yellow error bands around the expected limits. The purely theoretical cross section uncertainties on the signals are not considered when deriving limits and are not displayed in the plots. The effect of the beam-energy uncertainty on the observed limit is negligible and is not included.

While the limits displayed in Fig. 7.22 extend down to DM masses of 10 GeV, they can be extrapolated and applied to even lower masses, since both the change in cross section and acceptance is very small.

As discussed in detail in Chap. 6, the criteria of validity for an EFT are not met in all areas of the kinematic phase space that is probed. The EFT limits as presented here should be seen as benchmark scenarios. The approach of truncating the signal cross section according to the fraction of valid events that is presented in Chap. 6 is adapted and allows to judge the vulnerability of the limits to the validity problems.

The condition that the momentum transferred in the hard interaction, Q_{tr}, should be below the mediator particle mass: $Q_{tr} < M_{med}$, is used as a criterion of validity. A natural coupling strength of one and the maximal value that allows for a perturbative theory, 3.5, are used for the product of the mediator coupling to DM and Standard Model particles. Events are omitted if they do not fulfil the validity criterion, leading to a reduction of the signal cross section. This reduced cross section is then used to re-derive the limits on Λ, leading, after sufficient iterations, to the truncated limits presented in the limit plots in Fig. 7.22. Where the truncated limit lines are not drawn, no meaningful limit can be obtain when imposing the validity condition.

Assuming a thermal history (see Chap. 3 for details), the cross section bound can be related to the measured DM relic density (taken from Ref. [34]). The values of Λ and m_{DM} that correspond to the correct cross section to reproduce the relic density are denoted as a green line in Fig. 7.22. The calculation assumes that no other interaction than the one considered contributes to the production and annihilation of DM. If this assumption is justified and if DM is entirely made up of thermal relics, the region in parameter space where the cut-off scale limits exceed the relic density line are excluded. If a thermal relic DM candidate exists in these regions, additional annihilation mechanisms or operators need to be assumed to restore agreement with the relic density measurements.

Comparison to Direct Detection Within the effective field theory, the obtained collider bounds can be converted into limits on the DM-nucleon scattering cross section, which is probed by direct-detection (DD) experiments.[6] Details of this procedure are given in Sect. 3.6. The results are presented in Fig. 7.23.

Compared to the DD results, the collider bounds are especially competitive in the low-DM mass region: while DD experiments lose sensitivity at some point (the nuclear recoil would be too small to be detected) the collider performance is constant towards low m_{DM}. Different interaction operators translate to either spin-dependent (D8, D9) or spin-independent (C1, C5, D1, D5, D11) DM-nucleon scattering limits. While the spin-independent case is tightly constrained by direct detection experiments, the spin-dependent case is experimentally more challenging for this approach and affected by several theoretical uncertainties. The collider bounds perform equally well in both cases and outperform the DD bound over the whole DM mass range. Note that the above conclusions only hold within the specific assumptions made on both sides.

Comparison to Indirect Detection Similar to the direct detection case, the collider limits can be transformed into limits on the DM annihilation cross section, which is probed by indirect detection (ID) experiments (see Sect. 3.6). The results for the vector and axial-vector operators (D5 and D8) are compared to bounds from the gamma-ray telescopes Fermi-LAT [60] and HESS [61]. While the latter commonly assume Majorana DM, the EFT models used for the interpretation of this analysis employed Dirac DM. Therefore, the limits from the gamma-ray telescopes are scaled

---

[6] Note that the momentum transfer in DM-nucleon scatterings is very small ($\mathcal{O}(1\,\text{keV})$) which leaves the EFT approach fully applicable.

7.7 Interpretation

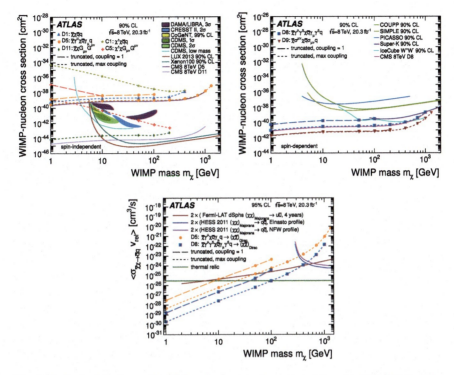

Fig. 7.23 Derived limits on the DM–nucleon scattering cross section (at 90% CL) for spin-independent (upper left) and spin-dependent interactions (upper right) as a function of the DM mass m_χ for different interaction operators. Results from direct-detection experiments for the spin-independent [48–54] and spin-dependent [55–59] cross section, and the CMS (not rescaled according to EFT validity) results [6] are compared to the limits provided by this analysis. Derived limits on the DM annihilation rate, defined as the product of the annihilation cross section σ and the relative DM velocity v, averaged over the velocity distribution ($\langle \sigma v \rangle$) at 95% CL as a function of DM mass are shown in the bottom plot. The results are presented for the operators D5 (vector interaction) and D8 (axial-vector interaction). For comparison, results from gamma-ray telescopes [60, 61] and the thermal relic density annihilation rate [47, 62] are displayed

up by a factor of two to allow for a direct comparison. In addition, the annihilation cross section corresponding to the measured relic abundance is displayed. While for high DM masses the telescopes are more powerful, collider limits can provide important insight over a large range of lower DM masses.

In all panels of Fig. 7.23 the impact of the questionable EFT validity is illustrated by the inclusion of the truncated limits in the comparison. The conclusion depends strongly on the considered operator and the coupling choice. Generally, the limits remain unchanged up to DM masses of $\mathcal{O}(100)$ GeV. The variation of the coupling strengths significantly affects the rescaled cross section limits, up to one order of magnitude.

7.7.2.2 Interpretation in terms of a Simplified Model

The problem of limited validity of the EFT models demands taking a step towards less general but more concrete models: so-called Simplified Models. Viewed from the EFT perspective, the mediator that is integrated out to obtain the effective operators is re-introduced: it has a propagator, can be produced on-shell and exhibits a finite width. For the interpretation of the present results, a Simplified Model assuming a Z'-like boson as the mediating particle between DM and the Standard Model is considered. A more detailed and comprehensive study of reinterpretations of DM collider results in terms of Simplified Models is presented in Chap. 8.

The Z'-like Simplified Model corresponds to the effective vector operator D5. Whereas in the EFT only the DM mass and the cut-off scale feature as model parameters, now the DM mass, the mediator mass, its width and two coupling strengths–the one between the mediator and the quarks and the one between the mediator and the DM particles–need to be specified. The obtained EFT limit and the deviation from it in case the Simplified Model is used can be seen in Fig. 7.24. For a fixed DM mass and mediator width, the observed 95% CL limit on the cross section constrains the coupling product, $\sqrt{g_q g_\chi}$. This is translated into an EFT-like cut-off scale via $\Lambda = M_{\text{med}}/\sqrt{g_q g_\chi}$ and presented as a function of the mediator mass. While the limits from EFT and Simplified Model agree above a mediator mass of 5 TeV, the EFT limits are significantly weaker in the region where the mediator is produced resonantly (700 GeV $< M_{\text{med}} <$ 5 TeV) and over-optimistic in the off-shell regime, where $M_{\text{med}} < m_{\text{DM}}$.

Figure 7.24 also presents the observed 95% CL upper limits on the product of couplings of the Simplified Model vertices in the plane of mediator and DM mass (M_{med} versus m_χ). Within this model, the regions to the left of the relic density line lead to larger values of the relic density than measured and would require an additional annihilation mechanism.

7.8 Conclusions

Monojet-like final states present a unique way to search for physics beyond the Standard Model at colliders. Especially in view of effective and simplified DM models, this analysis can constrain the relevant parameters powerfully. The search performed on 20.3 fb^{-1} of pp collisions at $\sqrt{s} = 8$ TeV, recorded with ATLAS at the LHC, was presented. The data is found to agree with the background estimate in all signal regions.

The analysis introduced a veto on isolated tracks that allows to further reduce electroweak backgrounds. Furthermore, a dedicated optimisation for DM signals was performed, leading to the replacement of the restriction on the number of jets with a topological cut on the ratio of E_T^{miss} and leading jet p_T. The background contributions from electroweak processes–most importantly from the irreducible $Z(\to \nu\bar{\nu})$ background–are estimated from MC that is normalised in dedicated control regions. The observation is then extrapolated to the signal regions via a transfer factor,

7.8 Conclusions

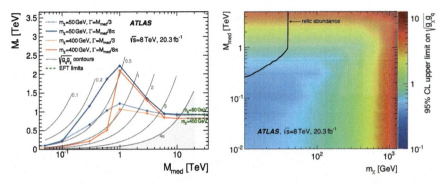

Fig. 7.24 Observed 95% CL limits obtained on a simplified Z'-like model are translated into limits on the suppression scale Λ (here denoted as M_*), shown as a function of the mediator mass M_{med} (left). DM masses of 50 GeV and 400 GeV and mediator widths of $M_{\mathrm{med}}/3$ and $M_{\mathrm{med}}/8\pi$ have been tested. The green, dashed lines at high mediator masses display the corresponding limits obtained from the EFT approach. The corresponding product of the coupling constants ($\sqrt{g_q\, g_\chi}$) is indicated by gray contours. Observed 95% CL upper limits on the product of couplings of the Simplified Model vertices in the plane of mediator and DM mass (M_{med} versus m_χ) are shown in the right plot. Parameters providing agreement with the measured relic abundance [47] are indicated by the black solid line

including the obtained normalisation, different acceptances of the processes and data-MC differences in efficiencies. This procedure allows to reduce the total systematic uncertainty on the background estimate to 2.7–14%.

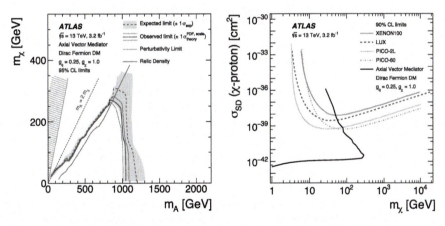

Fig. 7.25 Left: exclusion contours in the (m_{DM}, M_{med}) plane, obtained at 95% CL. The solid (dashed) curve shows the median of the observed (expected) limit. The red dotted band indicates the $\pm 1\sigma$ theory uncertainty on the observed limit, the yellow band shows the $\pm 1\sigma$ uncertainty on the expected limit. The red curve indicates the DM relic density. Right: the limits are compared to direct detection constraints on the spin-dependent DM–proton scattering cross section at 90% CL from the XENON100, LUX, and PICO experiments. Figure from Ref. [63]

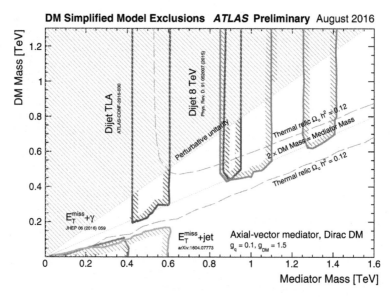

Fig. 7.26 95% CL exclusions by a selection of ATLAS Dark Matter searches on the ($m_{\rm DM}$, $M_{\rm med}$) plane. The results are obtained from analyses of 13 TeV pp collisions collected until summer 2016 (except for the limits labelled "Dijet 8 TeV" which are based on the 2012 8 TeV dataset). The Simplified Model of a leptophobic axial-vector mediator is considered. The exclusions are computed for a Dark Matter coupling $g_\chi = 1.5$ and a quark coupling $g_q = 0.1$ universal to all flavours. Dashed curves indicate signal points that are consistent with the DM relic density. A dotted curve indicates the region where on-shell mediator decays to DM are possible. Points in the plane where the model is in tension with the perturbative unitary considerations are indicated by the shaded triangle at the upper left. The exclusion regions, relic density contours, and unitarity curve are not applicable to other choices of coupling values or model. Figure from Ref. [65]

The findings are interpreted in terms of an effective field theory approach to DM pair production. The impact of its limited validity was explicitly included in the interpretation and presented along with the resulting bound on the cut-off scale. These results are compared to limits from direct and indirect detection experiments, revealing that, under the specified assumptions, collider searches can especially restrict low DM masses. As a step beyond the EFT, a Z'-like Simplified DM Model is studied as well. The EFT limits are comparable to the Simplified Model results only for very high mediator masses. In the resonance region, the limits get underestimated by the EFT, in the off-shell regime they are over-estimated. The limits on the coupling product was presented in the ($m_{\rm DM}$, $M_{\rm med}$) plane for this Simplified Model. While the inclusion of a Simplified-Model interpretation already allowed for interesting conclusions, a more comprehensive study of Simplified Models with a wider range of parameters, as presented in Chap. 8, needs to follow.

In the meanwhile, this analysis was performed also on the first 3.2 fb^{-1} of 13 TeV pp collision data [63]. A Simplified Model with an axial-vector mediator is considered in the interpretation (see Fig. 7.25). Also, with the rise of Simplified Models, the direct searches for the mediator particles, such as the dijet resonance

7.8 Conclusions

search is included in the DM picture. Figure 7.26 shows the complementarity of such searches and the missing transverse energy based *mono-X* approaches.[7]

Apart from additional statistics, the addition of shape information could improve the results of the analysis further, especially for signals with lower E_T^miss. Also, it might be beneficial to combine several mono-X channels in one analysis: CMS for example treats mono-jet and mono-W/Z final states, where the vector bosons decay hadronically, on equal footing appreciating that both signal scenarios can be constrained by both analysis channels [64]. This might become also relevant for final states of DM and top quarks which are discussed in detail in Chap. 9: if the tops are highly boosted, they are contained in large-radius jets and the overlap between a monojet-like scenario and a final state of DM and top quarks might become relevant.

References

1. CDF Collaboration, T. Aaltonen et al., A Search for dark matter in events with one jet and missing transverse energy in $p\bar{p}$ collisions at $\sqrt{s} = 1.96$ TeV. Phys. Rev. Lett. **108**, 211804 (2012). https://doi.org/10.1103/PhysRevLett.108.211804, arXiv:1203.0742 [hep-ex]
2. CMS Collaboration, S. Chatrchyan et al., Search for dark matter and large extra dimensions in monojet events in pp collisions at $\sqrt{s} = 7$ TeV. JHEP **09**, 094 (2012). https://doi.org/10.1007/JHEP09(2012)094, arXiv:1206.5663 [hep-ex]
3. CMS Collaboration, S. Chatrchyan et al., Search for dark matter and large extra dimensions in pp collisions yielding a photon and missing transverse energy. Phys. Rev. Lett. **108**, 261803 (2012). https://doi.org/10.1103/PhysRevLett.108.261803, arXiv:1204.0821 [hep-ex]
4. ATLAS Collaboration, Search for dark matter candidates and large extra dimensions in events with a jet and missing transverse momentum with the ATLAS detector. JHEP **1304**, 075 (2013). https://doi.org/10.1007/JHEP04(2013)075, arXiv:1210.4491 [hep-ex]
5. ATLAS Collaboration, G. Aad et al., Search for dark matter candidates and large extra dimensions in events with a photon and missing transverse momentum in pp collision data at $\sqrt{s} = 7$ TeV with the ATLAS detector. Phys. Rev. Lett. **110**(1), 011802 (2013). https://doi.org/10.1103/PhysRevLett.110.011802, arXiv:1209.4625 [hep-ex]
6. CMS Collaboration, V. Khachatryan et al., Search for dark matter, extra dimensions, and unparticles in monojet events in proton-proton collisions at $\sqrt{s} = 8$ TeV. Eur. Phys. J. **C75**(5), 235 (2015). https://doi.org/10.1140/epjc/s10052-015-3451-4, arXiv:1408.3583 [hep-ex]
7. ATLAS Collaboration, G. Aad et al., Search for pair-produced third-generation squarks decaying via charm quarks or in compressed supersymmetric scenarios in pp collisions at $\sqrt{s} = 8$ TeV with the ATLAS detector. Phys. Rev. **D90**(5), 052008 (2014). https://doi.org/10.1103/PhysRevD.90.052008, arXiv:1407.0608 [hep-ex]
8. ATLAS Collaboration, G. Aad et al., Search for dark matter in events with heavy quarks and missing transverse momentum in pp collisions with the ATLAS detector. Eur. Phys. J. **C75**(2), 92 (2015). https://doi.org/10.1140/epjc/s10052-015-3306-z, arXiv:1410.4031 [hep-ex]
9. ATLAS Collaboration, G. Aad et al., Search for dark matter in events with a Z boson and missing transverse momentum in pp collisions at $\sqrt{s} = 8$ TeV with the ATLAS detector. Phys. Rev. **D90**(1), 012004 (2014). https://doi.org/10.1103/PhysRevD.90.012004, arXiv:1404.0051 [hep-ex]

---

[7]Note that the complementarity between dijet and mono-X searches depends strongly on the coupling choice: increasing the mediator coupling to DM make mono-X searches more powerful, dijet searches become stronger when the mediator coupling to quarks is enhanced.

10. ATLAS Collaboration, G. Aad et al., Search for dark matter in events with a hadronically decaying W or Z boson and missing transverse momentum in pp collisions at $\sqrt{s} = 8$ TeV with the ATLAS detector. Phys. Rev. Lett. **112**(4), 041802 (2014). https://doi.org/10.1103/PhysRevLett.112.041802, arXiv:1309.4017 [hep-ex]
11. CMS Collaboration, V. Khachatryan et al., Search for physics beyond the standard model in final states with a lepton and missing transverse energy in proton-proton collisions at $\sqrt{s} = 8$ TeV. Phys. Rev. **D91**(9), 092005 (2015). https://doi.org/10.1103/PhysRevD.91.092005, arXiv:1408.2745 [hep-ex]
12. ATLAS Collaboration, G. Aad et al., Search for new phenomena in final states with an energetic jet and large missing transverse momentum in pp collisions at $\sqrt{s} = 8$ TeV with the ATLAS detector. Eur. Phys. J. **C75**(7), 299 (2015). https://doi.org/10.1140/epjc/s10052-015-3517-3, https://doi.org/10.1140/epjc/s10052-015-3639-7, arXiv:1502.01518 [hep-ex]
13. ATLAS Collaboration, M. Aaboud et al., Luminosity determination in pp collisions at $\sqrt{s} = 8$ TeV using the ATLAS detector at the LHC, arXiv:1608.03953 [hep-ex]
14. ATLAS Collaboration, G. Aad et al., Improved luminosity determination in pp collisions at $\sqrt{s} = 7$ TeV using the ATLAS detector at the LHC. Eur. Phys. J. **C73**(8), 2518 (2013). https://doi.org/10.1140/epjc/s10052-013-2518-3, arXiv:1302.4393 [hep-ex]
15. T. Gleisberg, S. Hoeche, F. Krauss, M. Schonherr, S. Schumann, F. Siegert, J. Winter, Event generation with SHERPA 1.1. JHEP **02**, 007 (2009). https://doi.org/10.1088/1126-6708/2009/02/007, arXiv:0811.4622 [hep-ph]
16. H.-L. Lai, M. Guzzi, J. Huston, Z. Li, P.M. Nadolsky, J. Pumplin, C.P. Yuan, New parton distributions for collider physics. Phys. Rev. **D82**, 074024 (2010). https://doi.org/10.1103/PhysRevD.82.074024, arXiv:1007.2241 [hep-ph]
17. M. Mangano et al., Alpgen, a generator for hard multiparton processes in hadronic collisions. JHEP **07**, 001 (2003), arXiv:hep-ph/0206293
18. G. Marchesini, B.R. Webber, G. Abbiendi, I.G. Knowles, M.H. Seymour, L. Stanco, HERWIG: A Monte Carlo event generator for simulating hadron emission reactions with interfering gluons. Version 5.1 - April 1991. Comput. Phys. Commun. **67**, 465–508 (1992). https://doi.org/10.1016/0010-4655(92)90055-4
19. G. Corcella, I.G. Knowles, G. Marchesini, S. Moretti, K. Odagiri, P. Richardson, M.H. Seymour, B.R. Webber, HERWIG 6.5 release note, arXiv:hep-ph/0210213 [hep-ph]
20. J.M. Butterworth, J.R. Forshaw, M.H. Seymour, Multiparton interactions in photoproduction at HERA. Z. Phys. **C72**, 637–646 (1996). https://doi.org/10.1007/BF02909195, 10.1007/s002880050286, arXiv:hep-ph/9601371 [hep-ph]
21. J. Pumplin, D.R. Stump, J. Huston, H.L. Lai, P.M. Nadolsky, W.K. Tung, New generation of parton distributions with uncertainties from global QCD analysis. JHEP **0207**, 012 (2002). https://doi.org/10.1088/1126-6708/2002/07/012, arXiv:hep-ph/0201195 [hep-ph]
22. Y.A. Golfand, E.P. Likhtman, Extension of the algebra of poincare group generators and violation of p invariance. JETP Lett. **13**, 323–326 (1971)
23. A.D. Martin, W.J. Stirling, R.S. Thorne, G. Watt, Parton distributions for the LHC. Eur. Phys. J. **C63**, 189–285 (2009). https://doi.org/10.1140/epjc/s10052-009-1072-5, arXiv:0901.0002 [hep-ph]
24. S. Frixione, B. Webber, Matching NLO QCD computations and parton shower simulations. JHEP **06**, 029 (2002), arXiv:hep-ph/0204244
25. S. Frixione, P. Nason, B.R. Webber, Matching NLO QCD and parton showers in heavy flavour production. JHEP **08**, 007 (2003), arXiv:hep-ph/0305252
26. B.P. Kersevan, E. Richter-Was, The Monte Carlo event generator AcerMC version 1.0 with interfaces to PYTHIA 6.2 and HERWIG 6.3. Comput. Phys. Commun. **149**, 142–194 (2003). https://doi.org/10.1016/S0010-4655(02)00592-1, arXiv:hep-ph/0201302 [hep-ph]
27. M. Czakon, P. Fiedler, A. Mitov, Total top-quark pair-production cross section at hadron colliders through $O(\alpha_S^4)$. Phys. Rev. Lett. **110**, 252004 (2013). https://doi.org/10.1103/PhysRevLett.110.252004, arXiv:1303.6254 [hep-ph]
28. N. Kidonakis, Two-loop soft anomalous dimensions for single top quark associated production with a W- or H-. Phys. Rev. **D82**, 054018 (2010). https://doi.org/10.1103/PhysRevD.82.054018, arXiv:1005.4451 [hep-ph]

29. M. Czakon, A. Mitov, Top++: a program for the calculation of the top-pair cross-section at hadron colliders. Comput. Phys. Commun. **185**, 2930 (2014). https://doi.org/10.1016/j.cpc.2014.06.021, arXiv:1112.5675 [hep-ph]
30. ATLAS Collaboration, ATLAS tunes of PYTHIA 6 and Pythia 8 for MC11, http://cds.cern.ch/record/1363300
31. T. Sjöstrand, S. Mrenna, P.Z. Skands, A brief introduction to PYTHIA 8.1. Comput. Phys. Commun. **178**, 852 (2008). https://doi.org/10.1016/j.cpc.2008.01.036, arXiv:0710.3820 [hep-ph]
32. J.M. Campbell, R.K. Ellis, C. Williams, Vector boson pair production at the LHC. JHEP **07**, 018 (2011), arXiv:1105.0020 [hep-ph]
33. J. Alwall, M. Herquet, F. Maltoni, O. Mattelaer, T. Stelzer, MadGraph 5: going beyond. JHEP **06**, 128 (2011). https://doi.org/10.1007/JHEP06(2011)128, arXiv:1106.0522 [hep-ph]
34. J. Goodman, M. Ibe, A. Rajaraman, W. Shepherd, T.M.P. Tait, H.-B. Yu, Constraints on dark matter from colliders. Phys. Rev. **D82**, 116010 (2010). https://doi.org/10.1103/PhysRevD.82.116010, arXiv:1008.1783 [hep-ph]
35. T. Sjostrand, S. Mrenna, P.Z. Skands, PYTHIA 6.4 physics and manual. JHEP **05**, 026 (2006). https://doi.org/10.1088/1126-6708/2006/05/026, arXiv:hep-ph/0603175 [hep-ph]
36. M.L. Mangano, M. Moretti, F. Piccinini, M. Treccani, Matching matrix elements and shower evolution for top-quark production in hadronic collisions. JHEP **01**, 013 (2007), arXiv:hep-ph/0611129 [hep-ph]
37. ATLAS Collaboration, G. Aad et al., Electron reconstruction and identification efficiency measurements with the ATLAS detector using the 2011 LHC proton-proton collision data. Eur. Phys. J. **C74**(7), 2941 (2014). https://doi.org/10.1140/epjc/s10052-014-2941-0, arXiv:1404.2240 [hep-ex]
38. ATLAS Collaboration, G. Aad et al., Characterisation and mitigation of beam-induced backgrounds observed in the ATLAS detector during the 2011 proton-proton run. JINST **8**, P07004 (2013). https://doi.org/10.1088/1748-0221/8/07/P07004, arXiv:1303.0223 [hep-ex]
39. ATLAS Collaboration, Selection of jets produced in proton-proton collisions with the ATLAS detector using 2011 data. Technical Report, ATLAS-CONF-2012-020, CERN, Geneva, Mar 2012, http://cds.cern.ch/record/1430034
40. G. Punzi, Sensitivity of searches for new signals and its optimization. in *eConf*, C030908, MODT002 (2003), arXiv:physics/0308063 [physics]
41. L. Lyons, D. Gibaut, P. Clifford, How to combine correlated estimates of a single physical quantity. Nucl. Instrum. Meth. **A270**, 110 (1988). https://doi.org/10.1016/0168-9002(88)90018-6
42. ATLAS Collaboration, G. Aad et al., Search for squarks and gluinos with the ATLAS detector in final states with jets and missing transverse momentum using 4.7 fb^{-1} of $\sqrt{s} = 7$ TeV proton-proton collision data. Phys. Rev. **D87**(1), 012008 (2013). https://doi.org/10.1103/PhysRevD.87.012008, arXiv:1208.0949 [hep-ex]
43. A. Denner, S. Dittmaier, T. Kasprzik, A. Mück, Electroweak corrections to monojet production at the LHC. Eur. Phys. J. **C73**(2), 2297 (2013). https://doi.org/10.1140/epjc/s10052-013-2297-x, arXiv:1211.5078 [hep-ph]
44. A. Denner, S. Dittmaier, T. Kasprzik, A. Muck, Electroweak corrections to dilepton + jet production at hadron colliders. JHEP **06**, 069 (2011). https://doi.org/10.1007/JHEP06(2011)069, arXiv:1103.0914 [hep-ph]
45. A. Denner, S. Dittmaier, T. Kasprzik, A. Muck, Electroweak corrections to W + jet hadroproduction including leptonic W-boson decays. JHEP **08**, 075 (2009). https://doi.org/10.1088/1126-6708/2009/08/075, arXiv:0906.1656 [hep-ph]
46. A.L. Read, Presentation of search results: the CL(s) technique. J. Phys. **G28**, 2693–2704 (2002). https://doi.org/10.1088/0954-3899/28/10/313
47. WMAP Collaboration, G. Hinshaw et al., Nine-year Wilkinson microwave anisotropy probe (WMAP) observations: cosmological parameter results. Astrophys. J. Suppl. **208**, 19 (2013). https://doi.org/10.1088/0067-0049/208/2/19, arXiv:1212.5226 [astro-ph.CO]
48. G. Angloher, M. Bauer, I. Bavykina, A. Bento, C. Bucci et al., Results from 730 kg days of the CRESST-II dark matter search. Eur. Phys. J. **C72**, 1971 (2012). https://doi.org/10.1140/epjc/s10052-012-1971-8, arXiv:1109.0702 [astro-ph.CO]

49. LUX Collaboration, D.S. Akerib et al., First results from the LUX dark matter experiment at the Sanford underground research facility. Phys. Rev. Lett. **112**, 091303 (2014). https://doi.org/10.1103/PhysRevLett.112.091303, arXiv:1310.8214 [astro-ph.CO]
50. CDMS Collaboration, R. Agnese et al., Silicon detector dark matter results from the final exposure of CDMS II. Phys. Rev. Lett. **111**, 251301 (2013). https://doi.org/10.1103/PhysRevLett.111.251301, arXiv:1304.4279 [hep-ex]
51. SuperCDMS Collaboration, R. Agnese et al., Search for low-mass weakly interacting massive particles with SuperCDMS. Phys. Rev. Lett. **112**(24), 241302 (2014). https://doi.org/10.1103/PhysRevLett.112.241302, arXiv:1402.7137 [hep-ex]
52. C. Aalseth, P. Barbeau, J. Colaresi, J. Collar, J.D. Leon, et al., Maximum likelihood signal extraction method applied to 3.4 years of CoGeNT data, arXiv:1401.6234 [astro-ph.CO]
53. DAMA Collaboration, R. Bernabei et al., First results from DAMA/LIBRA and the combined results with DAMA/NaI. Eur. Phys. J. **C56**, 333–355 (2008). https://doi.org/10.1140/epjc/s10052-008-0662-y, arXiv:0804.2741 [astro-ph]
54. XENON100 Collaboration, E. Aprile et al., Limits on spin-dependent WIMP-nucleon cross sections from 225 live days of XENON100 data. Phys. Rev. Lett. **111**(2), 021301 (2013). https://doi.org/10.1103/PhysRevLett.111.021301, arXiv:1301.6620 [astro-ph.CO]
55. PICASSO Collaboration, S. Archambault et al., Constraints on low-mass WIMP interactions on ^{19}F from PICASSO. Phys. Lett. **B711**, 153–161 (2012). https://doi.org/10.1016/j.physletb.2012.03.078, arXiv:1202.1240 [hep-ex]
56. Super-Kamiokande Collaboration, S. Desai et al., Search for dark matter WIMPs using upward through-going muons in Super-Kamiokande. Phys. Rev. **D70**, 083523 (2004). https://doi.org/10.1103/PhysRevD.70.083523, https://doi.org/10.1103/10.1103/PhysRevD.70.109901, arXiv:hep-ex/0404025 [hep-ex]
57. IceCube Collaboration, R. Abbasi et al., Limits on a muon flux from neutralino annihilations in the Sun with the IceCube 22-string detector. Phys. Rev. Lett. **102**, 201302 (2009). https://doi.org/10.1103/PhysRevLett.102.201302, arXiv:0902.2460 [astro-ph.CO]
58. E. Behnke et al., Improved limits on spin-dependent WIMP-proton interactions from a two liter CF_3I Bubble Chamber. Phys. Rev. Lett. **106**, 021303 (2011). https://doi.org/10.1103/PhysRevLett.106.021303, arXiv:1008.3518 [astro-ph.CO]
59. M. Felizardo et al., Final analysis and results of the phase II SIMPLE dark matter search. Phys. Rev. Lett. **108**, 201302 (2012). https://doi.org/10.1103/PhysRevLett.108.201302, arXiv:1106.3014 [astro-ph.CO]
60. Fermi-LAT Collaboration, M. Ackermann et al., Dark matter constraints from observations of 25 Milky Way satellite galaxies with the fermi large area telescope. Phys. Rev. **D89**, 042001 (2014). https://doi.org/10.1103/PhysRevD.89.042001, arXiv:1310.0828 [astro-ph.HE]
61. H.E.S.S. Collaboration, A. Abramowski et al., Search for a dark matter annihilation signal from the Galactic Center halo with H.E.S.S. Phys. Rev. Lett. **106**, 161301 (2011). https://doi.org/10.1103/PhysRevLett.106.161301, arXiv:1103.3266 [astro-ph.HE]
62. Planck Collaboration, P.A.R. Ade et al., Planck 2013 results. XVI. Cosmological parameters. Astron. Astrophys. **571**, A16 (2014). https://doi.org/10.1051/0004-6361/201321591, arXiv:1303.5076 [astro-ph.CO]
63. ATLAS Collaboration, M. Aaboud et al., Search for new phenomena in final states with an energetic jet and large missing transverse momentum in pp collisions at $\sqrt{s} = 13$ TeV using the ATLAS detector. Phys. Rev. **D94**(3), 032005 (2016). https://doi.org/10.1103/PhysRevD.94.032005, arXiv:1604.07773 [hep-ex]
64. CMS Collaboration, Search for dark matter in final states with an energetic jet, or a hadronically decaying W or Z boson using 12.9 fb^{-1} of data at $\sqrt{s} = 13$ TeV. Technical Report, CMS-PAS-EXO-16-037, CERN, Geneva (2016), https://cds.cern.ch/record/2205746
65. ATLAS Collaboration, DM Simplified Model Exclusion (2016), https://atlas.web.cern.ch/Atlas/GROUPS/PHYSICS/CombinedSummaryPlots/EXOTICS/ATLAS_DarkMatter_Summary_0simple/ATLAS_DarkMatter_Summary_0simple.pdf

Chapter 8
Constraints on Simplified Dark Matter Models from Mono-X Searches

Collider searches for Dark Matter (DM) in events with large missing transverse energy (E_T^{miss}) have been commonly interpreted within effective field theory (EFT) models of DM pair production. The advantage of this approach is that limits derived in terms of an EFT are applicable to a broad range of complete theories and depend only on the specification of a few parameters, namely the cutoff scale, Λ, and the DM mass, m_χ [1]. However, if the mass of the mediating particle is not significantly larger than the momentum transferred in a given interaction, which can be the case in pp collisions at the LHC, the EFT approach is no longer valid and constraints can differ significantly from those of an associated high-energy completion [1–5]. This has been discussed in detail in Chap. 6. In light of this and similar studies [6–9], Simplified Models emerge as the standard way in which LHC limits on DM are interpreted.

A Simplified Model is constructed by reintroducing the mediator between Standard Model and DM particles which was integrated out in the EFT approach. Like an EFT, a Simplified Model allows for comparisons of results obtained in the different fields of DM searches [10, 11]. It is defined by a set of parameters that is larger than the one of an EFT but still much smaller than for a full theory–namely the mass of the DM particle, m_χ, the mass of the mediator M_{med}, and the quark–mediator and DM–mediator coupling strengths, g_q and g_χ (or $g_{q\chi}$ in the case of a single, quark–DM–mediator coupling). Unlike in the case of EFTs, constraints calculated within the context of a Simplified Model are valid across a broad energy range.

In Chap. 7, a first example towards the use of Simplified Models has been presented. The search for new physics in events with large E_T^{miss} and an energetic jet is interpreted in terms of a Z'-like model, alongside with the EFT bounds. The motivation for the study presented in the following was to extend such limits to an enlarged parameter space of mediator and DM masses and the relative strength of the couplings to the visible and dark sectors. Apart from the Simplified Model studied in Chap. 7, a model describing an axial-vector interaction is considered, as well as a scalar-mediator exchange in the t-channel. Furthermore, in contrast to the limits on the Simplified Model presented in Chap. 7, the mediator width is considered to take the minimal value arising, given the model parameters, and not as a fixed value. The study compares the performance of three search channels. Specifically, searches where either a parton (manifesting in the detector as a narrow-radius jet), a leptonically-decaying Z boson, or a hadronically-decaying W or Z boson, reconstructed as a large-radius jet, is detected in addition to large E_T^{miss} are examined. In addition, constraints from the relic density and from direct detection experiments are considered. This study was developed and performed by me and three other authors. It was published in the beginning of 2016 [12].

After the introduction of the phenomenologically distinct set of Simplified Models chosen for this study in Sect. 8.1, the recast of the experimental searches is discussed in Sect. 8.2. The results are presented in Sect. 8.3, followed by concluding remarks (Sect. 8.4).

8.1 Simplified Models of Dark Matter Production

In the following, it is assumed that the DM particle, χ, is a weakly interacting Dirac fermion, a singlet under the Standard Model gauge group and the lightest stable new particle. It is assumed to couple to Standard Model quarks via a *mediator* particle.[1]

The s-channel models chosen for this analysis are Z'-type models characterised by vector (sV) or axial-vector (sA) couplings between the dark and Standard Model sector. The mediator is an Standard Model-singlet vector particle, denoted ξ. Such models have been also studied earlier [14–25]. They are described by the following interaction Lagrangians:

$$\mathcal{L}_{sV} \supset \xi_\mu \left[\sum_q g_q \bar{q} \gamma^\mu q + g_\chi \bar{\chi} \gamma^\mu \chi \right], \tag{8.1}$$

$$\mathcal{L}_{sA} \supset \xi_\mu \left[\sum_q g_q \bar{q} \gamma^\mu \gamma_5 q + g_\chi \bar{\chi} \gamma^\mu \gamma_5 \chi \right], \tag{8.2}$$

where the sum is taken over all quarks.

The t-channel model (abbreviated tS) is motivated by analogy with a common supersymmetric scenario in which neutralino DM interacts with the Standard Model sector via t-channel exchange of a squark [26]. The mediator is a scalar particle which is necessarily charged and coloured (labelled as ϕ). Such a model has been studied in the context of the LHC by a number of groups [10, 23, 27–33]. Note that in a supersymmetric scenario the DM particle is a Majorana fermion. The collider phenomenology of a Majorana fermion DM particle is kinematically identical to

---
[1]Couplings between the mediator and Standard Model leptons or gluons are generally possible and have been studied (e.g. [3, 13]) but are not considered here.

8.1 Simplified Models of Dark Matter Production

the corresponding Dirac case, the cross section varies by a simple factor. Hence the results for Dirac fermion DM can be easily transferred to the Majorana case.[2]

The t-channel mediator is allowed to couple to either left or right-handed quarks, since it could be an SU(2) doublet or singlet, respectively. Since the LHC is insensitive to the chirality of the quarks, it is assumed for simplicity that ϕ couples to left-handed quarks only, and is an $SU(2)$ doublet, allowing for the radiation of a W boson off the mediator. In order to respect minimal flavour violation, three generations of mediator doublets ϕ_i, with equal masses and couplings are included. The interaction Lagrangian for this model is then given by:

$$\mathcal{L}_{tS} \supset \sum_i g_{q\chi} \bar{Q}_i P_R \phi_i \chi + \text{h.c.}, \tag{8.3}$$

where the sum is taken over the three quark doublets, $g_{q\chi}$ is the DM–quark–mediator coupling (equal for each generation), and P_R is the chiral projection operator.

8.1.1 Mono-X Signatures

Final states of large E_T^{miss} and one energetic object, commonly referred to as mono-X signatures, are interesting to search for new physics, in particular in view of DM models. Since DM particles are not expected to interact with the detector material, they appear as missing transverse energy when balanced against a visible object, X, that is radiated from the initial or intermediate state. For the s-channel Simplified Models (SiMs) discussed above, only initial-state radiation is possible, due to the fact that the mediator is taken to be a Standard Model singlet. Example diagrams can be found in Fig. 8.1a, b. For the tS model, radiation of a gluon or electroweak (EW) boson is permitted both from initial state partons (Fig. 8.1c) or from the charged mediator (Fig. 8.1d).

The production of a jet in association with the invisible χ pair is expected to have the highest cross section, due to the strong coupling and prevalence of quarks and gluons in the initial state. However, this might not be the case for some particular models, e.g. if very large couplings between mediator and Z-bosons are assumed. Also generally, it is interesting to consider other channels, since they present a unique experimental signature. Apart from the monojet scenario, two additional channels are included in the following discussion. First, the relatively clean and easy to reconstruct leptonically-decaying mono-$(Z \rightarrow \ell^+\ell^-)$ channel is considered (abbreviated as mono-Z(lep)). Second, one can profit from the large branching fraction of hadronic W/Z decays in the mono-$(W/Z \rightarrow jj)$ channel (abbreviated as mono-W/Z(had)),

---

[2]Only in the validation of the mono-Z(lep) channel Majorana DM is considered, see Sect. 8.2.4.

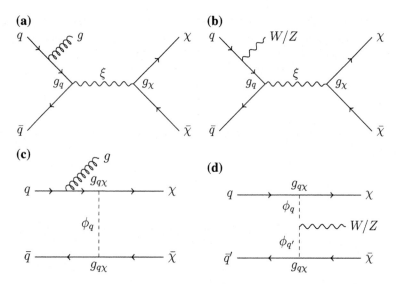

Fig. 8.1 Representative Dark Matter pair-production processes with a gluon or W/Z boson in the final state for the s-channel (**a**, **b**) and t-channel (**c**, **d**) models

Table 8.1 Mass and coupling points chosen for this analysis. Values in parantheses are only included in the mono-Z(lep) channel, where the faster event generation allowed for higher granularity. For the t-channel model, $M_{\text{med}} > m_\chi$ is required to ensure stability of the DM particle

m_χ [GeV]	M_{med} [GeV]	s-channel		t-channel
		g_q	g_χ	g_{qx}
1, (3), 10, (30), 100, (300), 1000	1, 2, 10, 20, 100, 200, 1000, 2000	1	0.2, 0.5, 1, 2, 5	1

reducing backgrounds via boson-tagging of large-radius jets.[3] In both cases, the large multi-jet background is efficiently reduced, and different experimental and theoretical uncertainties make these an interesting alternative to the monojet channel.

8.1.2 Mass and Coupling Points

A representative set of DM and mediator masses, listed in Table 8.1, are chosen for each detection channel. All (m_χ, M_{med}) combinations are allowed in the sV and sA models, while in the tS model M_{med} must be larger than m_χ to ensure the stability of the DM particle. The couplings g_q and g_{qx} are set to unity, while the DM-mediator

---

[3] One of the first Run II Dark Matter search results from ATLAS was from this channel [34], released during the preparation of this study.

8.1 Simplified Models of Dark Matter Production

coupling in the s-channel models, g_χ, is varied from 0.2 to 5. The mediator masses are chosen to cover a broad range of parameter space and to cover three regimes: (near-)degenerate ($M_\text{med} \approx m_\chi$), on-shell ($M_\text{med} \geq 2m_\chi$) and off-shell ($M_\text{med} < 2m_\chi$). For the couplings g_q and g_χ to remain within the perturbative regime, they are required to satisfy $g_q, g_\chi \leq 4\pi$, though stronger perturbativity requirements exist [7].

8.1.3 Mediator Width

When considering SiMs, it is important to ensure that the mediator width is treated appropriately, as it can impact both the cross section estimation and, in some cases, the kinematic behaviour of the signals.

In previous analyses (e.g in the *Monojet Analysis* presented in Chap. 7) mediators of fixed width ranging from $\Gamma = M/8\pi$ to $\Gamma = M/3$ have been commonly considered [35–37]. The smallest width, $\Gamma = M/8\pi$, corresponds to a mediator which couples only to one helicity and flavour of quarks with $g_q = 1$ [35]. This approach is motivated by the observation that in the case of a mediator exchanged in the s-channel and produced on-shell, the cross section at the resonance is maximally enhanced when Γ is small [35]. Hence, assuming a fixed, small width can be considered conservative. In this study, the mediator widths include couplings to all kinematically accessible quark flavours, the minimum width for each model is given by [38]:

$$\Gamma_{sV} = \frac{g_\chi^2 M}{12\pi}\left(1 + \frac{2m_\chi^2}{M^2}\right)\left(1 - \frac{4m_\chi^2}{M^2}\right)^{\frac{1}{2}} \Theta(M - 2m_\chi)$$

$$+ \sum_q \frac{g_q^2 M}{4\pi}\left(1 + \frac{2m_q^2}{M^2}\right)\left(1 - \frac{4m_q^2}{M^2}\right)^{\frac{1}{2}} \Theta(M - 2m_q), \quad (8.4)$$

$$\Gamma_{sA} = \frac{g_\chi^2 M}{12\pi}\left(1 - \frac{4m_\chi^2}{M^2}\right)^{\frac{3}{2}} \Theta(M - 2m_\chi)$$

$$+ \sum_q \frac{g_q^2 M}{4\pi}\left(1 - \frac{4m_q^2}{M^2}\right)^{\frac{3}{2}} \Theta(M - 2m_q), \quad (8.5)$$

$$\Gamma_{tS} = \sum_q \frac{g_{q\chi}^2 M}{16\pi}\left(1 - \frac{m_q^2}{M^2} - \frac{m_\chi^2}{M^2}\right)$$

$$\times \sqrt{\left(1 - \frac{m_q^2}{M^2} + \frac{m_\chi^2}{M^2}\right)^2 - 4\frac{m_\chi^2}{M^2}} \;\Theta(M - m_q - m_\chi). \quad (8.6)$$

Note that the mediator may decay to other Standard Model or new-physics particles [21], depending on the assumed underlying complete model, but this is not expected to have a large effect on the kinematic distributions as long as the width remains relatively small [38].

8.1.4 Rescaling Procedure

For each point in the (m_χ, M_{med}) plane, the model is constrained by placing a limit on the couplings ($\sqrt{g_q g_\chi}$ for the s-channel models and $g_{q\chi}$ for the t-channel model). This could be done by adding a third dimension to the (m_χ, M_{med}) grid of simulated signal samples, namely the varying couplings. However, this study exploits the fact that a change of the couplings (and consequently the width) of the mediator does not significantly change the kinematic behaviour of the model such that the cross section can be derived via a rescaling procedure considering the couplings and the width.

The fact that the mediator width (and therefore the couplings) do not greatly affect a model's kinematic behaviour (with the notable exception of the tS model in the monojet channel) is demonstrated in Fig. 8.2. The E_T^{miss} distribution, as a proxy for the full selection in each analysis, is plotted for the sV (representing both the sV and sA model) and tS model for two DM mass points and a demonstrative set of couplings such that $\Gamma < M_{\text{med}}/2$. The E_T^{miss} distribution is found to be mostly independent of the mediator width for the s-channel models in the monojet channel, and all models in the mono-Z(lep) channel.[4] However, there is a clear variation in the kinematic behaviour of the t-channel model in the monojet channel if the width is varied. This can be attributed to additional diagrams which are accessible only in this channel, featuring a gluon in the initial state and subsequently allowing the mediator to go on-shell. In this scenario, when the quark and DM masses are both small compared to the mediator mass, they equally share its energy, leading to a peak in the E_T^{miss} distribution at approximately half the mediator mass.

If the kinematic distribution is independent of the width, the impact of the selection cuts in each channel is assumed to be unchanged by the choice of couplings. In this case, the following relations approximately hold:

$$\sigma \propto \begin{cases} g_q^2 g_\chi^2 / \Gamma & \text{if } M_{\text{med}} \geq 2m_\chi \\ g_q^2 g_\chi^2 & \text{if } M_{\text{med}} < 2m_\chi \end{cases} \qquad (8.7)$$

in the sV and sA models [22], and:

$$\sigma \propto g_{q\chi}^4 \qquad (8.8)$$

---

[4]In this discussion, the mono-W/Z(had) channel can be assumed to follow the same logic as the mono-Z(lep) channel.

8.1 Simplified Models of Dark Matter Production

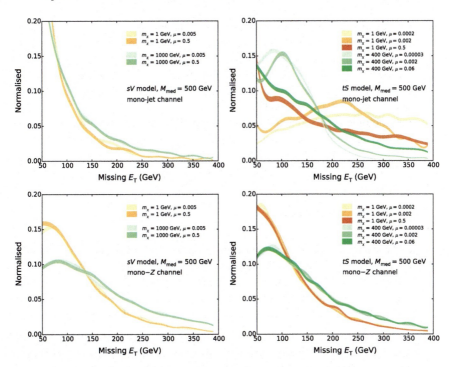

Fig. 8.2 The E_T^{miss} distribution of the sV and tS models in the monojet and mono-Z(lep) channels normalised to unity, for some exemplary masses. The line widths represent the statistical uncertainties. The parameter μ is defined as Γ/M_{med}, and is used to demonstrate the impact of a changing width. The tS model in the monojet channel shows a clear width-dependence

in the tS model. When valid, these approximations allow for a great simplification of the limit calculation, and for this reason, the primary results of this study are restricted to regions of parameter space where $\Gamma/M_{med} < 0.5$ such that the rescaling relations can be applied.

A full study of the tS model within the monojet channel, where altering the coupling can lead to changed kinematic behaviour, has been performed elsewhere [30], and requires the production of individual samples for each coupling point. It is not discussed further here.

8.2 Recasting Mono-X Constraints

The procedure for recasting existing mono-X analyses to obtain SiMs constraints follows a simple cut-and-count methodology. First, signal events are simulated (described below in Sect. 8.2.1) with object p_T smearing applied to approximate the detection efficiency, ϵ, of the ATLAS detector for which the DELPHES package [39] is used. The event selection criteria of the mono-X analysis of interest is then applied to the simulated signal samples. Events surviving the selection criteria are counted to determine the fraction of accepted signal events (referred to as the acceptance, $\mathcal{A}$). This is then used in combination with channel-specific model-independent limits on new physics events reported by the analyses to constrain the parameter space of the given model.

Monojet, leptonic mono-Z and hadronic mono-W/Z constraints are derived from ATLAS searches in these channels [37, 40, 41].[5] The relevant analysis details are described in Sects. 8.2.3, 8.2.4 and 8.2.5 respectively.

8.2.1 Signal Simulation

Monte Carlo simulated event samples are used to model the expected signal for each channel and for each SiM. Leading order matrix elements for the process $pp \to \chi\bar{\chi} + X$, where X is specifically one or two jets,[6] a $Z(\to \ell^+\ell^-)$ boson or a $W/Z(\to jj)$ boson, are first simulated using MADGRAPH5_aMC@NLO [42] with the MSTW2008lo68cl set of parton distribution functions (PDFs) [43]. The renormalisation and factorisation scales are set to the sum of $\sqrt{m^2 + p_T^2}$ for all particles in the final state. Showering and hadronisation are then performed by PYTHIA 8 [44] using the ATLAS underlying-event tune AU2-MSTW2008LO [45]. Jet reconstruction is performed by FASTJET [46] using the anti-k_T algorithm with a radius parameter of $R = 0.4$. Similarly, the reconstruction of large-radius jets for the mono-W/Z(had) channel is performed using the Cambridge-Aachen algorithm with $R = 1.2$. The latter also includes a mass-drop filtering procedure with mass drop $\mu = m_{sub-jet}/m_{jet}$ = 0.67 and $\sqrt{y} = 0.4$, where $\sqrt{y} = \min(p_{T_{j_1}}, p_{T_{j_2}})\Delta R/m_{jet}$ is the momentum balance of the two leading subjets (see Ref. [47] for further details). These requirements favour large-R jets with two balanced sub-jets, consistent with the decay of an electroweak boson to two close-by quarks. Lastly, the detector response is approximated by applying Gaussian smearing factors to the p_T of all leptons and jets using DELPHES.

8.2.1.1 Parton Matching Scheme

In the ATLAS monojet analysis [37] described in Chap. 7 and Sect. 8.2.3, matching of partons generated in MADGRAPH to jets generated in PYTHIA 8 is performed using the MLM scheme [48], with two matching scales per mass/coupling point. In combination, the matching scale values span a broad kinematic range with a cut placed on the leading jet p_T per event to avoid double-counting. This treatment aims

---

[5]The CMS collaboration published similar analyses with comparable results. Only the ATLAS set of results were recast for this study.

[6]Jets are seeded by any parton excluding the (anti-)top quark.

8.2 Recasting Mono-X Constraints

to mitigate the impact of the matching scale on the shape of the p_T and E_T^{miss} distributions, reducing the uncertainty in those areas of phase space where the transferred momentum is significantly larger or smaller than the matching scale. For the analysis recast presented here, a single matching scale of 80 GeV is used. Although not ideal, this approach suitably reproduces the results of the ATLAS monojet analysis for the masses of interest (see Sect. 8.2.3) while being less complex and computational expensive.

8.2.2 Limit Setting Strategy

8.2.2.1 Nominal Values

From each experimental analysis, the obtained model-independent upper limit on $\sigma \times \mathcal{A} \times \epsilon$ is taken. Together with the signal acceptance and efficiency, estimated from the simulated signal samples for each model and each set of parameters, the limit on the signal cross section, $\sigma_{\lim}$, is derived from the model-independent limits. Equation 8.7 is then used to convert $\sigma_{\lim}$ into a bound on the couplings. Exploiting the fact that in the s-channel on-shell case, the width can be expressed as a function of g_q and the ratio g_χ/g_q, the relations are as follows:

$$\sqrt{g_q g_\chi}_{\lim} = \begin{cases} \sqrt{g_q g_\chi}_{\text{gen}} \times \left(\sigma_{\lim}/\sigma_{\text{gen}}\right)^{\frac{1}{2}} & \text{if } M_{\text{med}} \geq 2m_\chi \ (s-\text{channel}) \\ \sqrt{g_q g_\chi}_{\text{gen}} \times \left(\sigma_{\lim}/\sigma_{\text{gen}}\right)^{\frac{1}{4}} & \text{if } M_{\text{med}} < 2m_\chi \end{cases} \quad (8.9)$$

where $\sqrt{g_q g_\chi}_{gen}$ and σ_{gen} are the input couplings and signal cross sections (taken from PYTHIA 8), respectively. For each model point, the signal region providing the best expected limit is chosen.

8.2.2.2 Uncertainty Estimation

As seen, both σ_{gen} and $\mathcal{A} \times \epsilon$ enter the calculation of the nominal limits and are hence subject to systematic uncertainties affecting signal cross section, acceptance and efficiency determined from the simulated signal samples. They are estimated by evaluating the three dominant sources of systematic uncertainties: the choice of factorisation and renormalisation scales, the assumed value for the strong coupling constant (α_s) and the choice of the used PDF set.

The impact of the factorisation and renormalisation scales are assessed by varying them simultaneously by factors of two and one half. It is assumed that the systematic uncertainty introduced by α_s at matrix-element level is negligible when compared to the differences between PDF sets, as demonstrated to be valid in Ref. [49]. The variation of α_s in the parton shower together with the change of PDF is realised

Table 8.2 The sources of systematic uncertainty considered in this analysis are summarised. The systematic uncertainty is taken from the resulting changes to the acceptance and cross section in comparison to their nominal values for each signal point

	Nominal	Variations
PDF/tune	MSTW2008lo68cl + AU2-MSTW2008LO	NNPDF2.1LO + Monash tune, CTEQ6L1 + AU2-CTEQ6L1
Factorisation and Renormalisation scales	$\times 1$	$\times 0.5, \times 2$
Matching scale (monojet only)	80 GeV	40 GeV, 160 GeV

by the use of specific tunes in PYTHIA 8 to estimate the uncertainty on σ_{gen}. The nominal choices of PDF and MC tune are varied to NNPDF2.1LO PDF + *Monash* tune [50], and to CTEQ6L1 PDF and ATLAS UE AU2-CTEQ6L1 tune. For the monojet channel, the impact of the matching scale is assessed in similar manner than the factorisation and renormalisation scales, namely the matching scale is varied by factors of two and one half. All sources of systematic uncertainty are summarised in Table 8.2.

The average variation in the nominal value of σ_{lim} resulting from each systematic source is added in quadrature and propagated to $\sqrt{g_q g_\chi}$ to obtain the total systematic uncertainty. This process is adjusted slightly to account for the inclusion of statistical uncertainties, which are estimated conservatively by taking the 95% CL *lower* limit on $\mathcal{A} \times \epsilon$ as suggested by the Wald approximation [51], i.e. $\mathcal{A} \times \epsilon \rightarrow (\mathcal{A} \times \epsilon) - \Delta(\mathcal{A} \times \epsilon)$.

8.2.3 Monojet Channel

The ATLAS monojet analysis [37] was designed to set limits on several new physics scenarios, including the production of DM via a set of effective operators. The analysis also includes a brief study of a Z' DM model which is analogous to the sV model discussed here. A detailed description of this analysis is given in Chap. 7.

At least one hard jet above $p_T^{j_1} > 120$ GeV and within $|\eta| < 2.0$ is required in the signal selection. Leading jet and E_T^{miss} have to satisfy $p_T^{j_1}/E_T^{\text{miss}} > 0.5$ to ensure a monojet-like topology (note that there is no upper limit placed on the number of jets per event). Events must then fulfil $|\Delta\phi(j, \vec{E}_T^{\text{miss}})| > 1.0$, where j is any jet with $p_T > 30$ GeV and $|\eta| < 4.5$. This criterion reduces the multijet background contribution for which the large E_T^{miss} originates mainly from jet energy mis-measurements. The contribution from the dominant background processes, $W/Z+$ jets, is reduced by applying a veto on events containing muons or electrons with $p_T > 7$ GeV. They are estimated in background-enriched control regions. Furthermore, the original analysis applied a veto on isolated tracks that reduces the electroweak backgrounds, especially those containing hadronically decaying tau leptons, which is not considered here.

8.2 Recasting Mono-X Constraints

Table 8.3 The ATLAS monojet E_T^{miss} signal regions and corresponding observed and expected model-independent upper limits on $\sigma \times \mathcal{A} \times \epsilon$ at 95% confidence level. (adapted from Ref. [37])

Signal region	E_T^{miss} threshold [GeV]	$\sigma \times \mathcal{A} \times \epsilon$ [fb] (obs)	$\sigma \times \mathcal{A} \times \epsilon$ [fb] (exp)
SR1	150	726	935
SR2	200	194	271
SR3	250	90	106
SR4	300	45	51
SR5	350	21	29
SR6	400	12	17
SR7	500	7.2	7.2
SR8	600	3.8	3.2
SR9	700	3.4	1.8

This means that the signal efficiency is expected to be slightly higher and the signal-to-background ratio is expected to be lower (around 10%) than in the ATLAS analysis. Nine separate signal regions are defined by increasing lower thresholds on E_T^{miss}, which range from 150 to 700 GeV as shown in Table 8.3.

The ATLAS analysis observed no significant deviation of observed events from the expected SM backgrounds. Subsequently, model-independent limits on new physics signatures were set on the visible cross section, $\sigma \times \mathcal{A} \times \epsilon$. These are listed in Table 8.3.

The signal simulation procedure outlined in Sect. 8.2.1 and the implementation of the selection criteria discussed above were validated for the monojet channel via the reproduction of the ATLAS limits on the EFT suppression scale, $\Lambda \equiv M_{\text{med}}/\sqrt{g_q g_\chi}$, for the Z' model.

A comparison of SR7 limits[7] for a representative sample of mediator masses with $m_\chi = 50$ GeV, $\Gamma = M/8\pi$ and $\sqrt{g_q g_\chi} = 1$ is presented in Table 8.4. An agreement of the obtained limits within about 12% for all samples is observed. A discrepancy of a few percent is expected given the differences in signal simulation: the simplified matching procedure discussed in Sect. 8.2.1.1 introduces an additional uncertainty of approximately 25% for events with $E_T^{\text{miss}} > 350$ GeV when compared to the approach utilised by the ATLAS monojet analysis. Further uncertainties are introduced by the jet smearing approximation used in place of a full detector simulation and by the 95% CL estimation procedure (outlined in Sect. 8.2.2) used instead of a thorough statistical treatment (e.g with HistFitter [52]). The results are consistently more conservative than those of the ATLAS analysis, so the approach is considered acceptable.

---

[7] This signal region is the only one for which the ATLAS analysis publicly provided the EFT limits.

Table 8.4 Comparison of the 95% CL upper limits on Λ from this work and from the ATLAS monojet analysis [37]. The limits are for an s-channel vector mediator model with $m_\chi = 50$ GeV and $\Gamma = M_{\text{med}}/8\pi$, and for the process $pp \rightarrow \chi\bar{\chi} + 1(2)$ jets with QCUT = 80 GeV. Note that Λ^{gen} is the input suppression scale

Λ^{gen} [TeV]	$\Lambda^{95\%\text{CL}}$ [GeV] (ATLAS)	$\Lambda^{95\%\text{CL}}$ [GeV] (this work)	Difference (%)
0.05	91	89	2.2
0.3	1151	1041	7.3
0.6	1868	1535	11.8
1	2225	1732	12.0
3	1349	1072	6.8
6	945	769	8.5
10	928	724	10.6
30	914	722	9.6

8.2.4 Mono-Z(lep) Channel

The ATLAS mono-Z(lep) analysis [40] was developed to constrain a set of EFT operators of DM production. It also includes a brief study of a t-channel SiM similar to the tS model.

The selection criteria for this analysis can be summarised as follows (see the ATLAS publication [40] for a full description). Electrons (muons) are required to have a $p_T > 20$ GeV, and $|\eta| < 2.47$ (2.5). Two opposite-sign, same-flavour leptons are selected, and required to have an invariant mass and pseudorapidity such that $m_{\ell\ell} \in [76, 106]$ GeV and $|\eta^{\ell\ell}| < 2.5$. The reconstructed Z boson should be approximately back-to-back and balanced against the E_T^{miss}, ensured with the selections $\Delta\phi(\vec{E}_T^{\text{miss}}, p_T^{\ell\ell}) > 2.5$ and $|p_T^{\ell\ell} - E_T^{\text{miss}}| / p_T^{\ell\ell} < 0.5$. Events containing a jet with $p_T > 25$ GeV and $|\eta| < 2.5$ are vetoed. Events are also vetoed if they contain a third lepton with $p_T > 7$ GeV. The signal regions are defined by increasing lower E_T^{miss} thresholds of $E_T^{\text{miss}} > 150, 250, 350, 450$ GeV.

A cut-and-count strategy is followed to estimate the expected SM backgrounds in each signal region. The limits on $\sigma \times \mathcal{A} \times \epsilon$ are not included in the published results, so the numbers of expected and observed events, along with the associated uncertainties, are used and converted into model-dependent upper limits with a simple implementation into the HistFitter framework [52] using a frequentist calculator and a one-sided profile likelihood test statistic. The results of this process are displayed in Table 8.5. Note that only the two signal regions with E_T^{miss} thresholds of 150 and 250 GeV are used here, as the applied simplified HistFitter approach is inadequate to handle the very low statistics of signal regions with higher E_T^{miss} thresholds. These upper limits are also used for the validation of the mono-Z(lep) signal generation and selection procedures.

The ATLAS mono-Z(lep) results include an upper limit on the coupling $g_{q\chi}$ for a t-channel SiM analogous to the tS model. This model is used to validate

8.2 Recasting Mono-X Constraints

Table 8.5 The ATLAS mono-Z(lep) + E_T^{miss} signal regions and corresponding observed and expected model-independent upper limits on $\sigma \times \mathcal{A} \times \epsilon$ at 95% confidence level. (adapted from Ref. [40], using HistFitter)

Signal region	E_T^{miss} threshold [GeV]	$\sigma \times \mathcal{A} \times \epsilon$ [fb] (obs)	$\sigma \times \mathcal{A} \times \epsilon$ [fb] (exp)
SR1	150	1.59	1.71
SR2	250	0.291	0.335

Table 8.6 Comparison of the 95% CL upper limit on $g_{q\chi}$ from this work and from the ATLAS mono-Z(lep) analysis [40]. The limits are for a variant of the t-channel scalar mediator model with Majorana Dark matter for the process $pp \to \chi\bar{\chi} + Z(\to e^+e^-/\mu^+\mu^-)$

m_χ [GeV]	M_{med} [GeV]	$g_{q\chi}^{95\%CL}$ (ATLAS)	$g_{q\chi}^{95\%CL}$ (this work)	Difference (%)
10	200	1.9	2.0	5.3
	500	2.8	3.2	14.3
	700	3.5	4.4	25.7
	1000	4.5	5.2	15.6
200	500	3.4	4.0	17.6
	700	4.2	4.5	7.1
	1000	5.2	5.3	1.9
400	500	5.5	5.7	3.6
	700	6.1	6.5	6.6
	1000	7.2	7.4	2.8
1000	1200	23.3	24.1	3.4

the signal generation and selection procedures. Note the following differences: the ATLAS model includes just two mediators (*up*- and *down*-type) where here six are considered, the DM particle is taken to be a Majorana fermion where here a Dirac particle is assumed, and the couplings $g_{t,b\chi}$ are set to zero while here universal couplings to all three quark generations are taken.

Table 8.6 shows the 95% CL upper limits on $g_{q\chi}$ that are calculated using the generation procedure described above and the values in Table 8.5, compared with the limits from the ATLAS analysis. Also shown is the difference as a percentage of the ATLAS limit. Reasonable agreement is observed: most of the 11 points in parameter space are within 10% of the ATLAS limits, and all are within 26%. Additionally, the results are consistently more conservative, which is to be expected given the differences in the generation procedure. As in the case of the monojet validation, further differences are expected from the use of p_T smearing applied to the leptons (rather than a full detector simulation) and from the simplified treatment of systematics, since $\sigma \times \mathcal{A} \times \epsilon$ is obtained independently.

8.2.5 Mono-W/Z(had) Channel

The ATLAS mono-W/Z(had) search [41] was optimised for a set of effective DM operators. Originally, it was designed to exploit the constructive interference of W boson emission from opposite-sign up-type and down-type quarks, leading to the mono-W channel being the dominant one for DM production. Recent studies [53] have revealed this scenario to violate gauge invariance and so it is not further discussed here.

The mono-W/Z(had) event selection is realised as follows. Large-radius (large-R) jets are selected using a mass-drop filtering procedure (see Sect. 8.2.1) to suppress non-W/Z processes. Events are required to contain at least one large-R jet with $p_T > 250$ GeV, $|\eta| < 1.2$ and a mass, m_{jet}, within a 30–40 GeV window of the W/Z mass (i.e. $m_{jet} \in [50, 120]$ GeV). In order to reduce the $t\bar{t}$ and multijet backgrounds, a veto removes events containing a small-R jet with $\Delta\phi(\text{jet}, E_T^{miss}) < 0.4$, or containing more than one small-R jet with $p_T > 40$ GeV, $|\eta| < 4.5$, and ΔR(small-R jet, large-R jet) > 0.9. Events containing electrons, muons and photons are vetoed if their p_T is larger than 10 GeV and if they are within $|\eta| < 2.47$ (electrons), 2.5 (muons), 2.37 (photons). Two signal regions were defined with $E_T^{miss} > 350$ GeV and $E_T^{miss} > 500$ GeV.

The ATLAS analysis used a shape fit of the large-R jet mass distribution to derive the limits on new physics in the two signal regions. The shapes are not taken into account in this study but the published number of expected and observed events in the signal regions are converted into upper limits on the expected and observed number of new physics events using the HistFitter framework, as is done in the mono-Z(lep) channel. For the $E_T^{miss} > 500$ GeV signal region, the obtained limits are shown in Table 8.7. This signal region was found to be optimal for most operators studied by the ATLAS analysis. The signal region with $E_T^{miss} > 350$ GeV is not considered here in the recasting procedure, since the cut-and-count limits extracted could not be convincingly validated. With this lower E_T^{miss} threshold the differences between a shape fit and a cut-and-count result are expectedly more severe.

The event generation and selection procedures for the mono-W/Z(had) channel is validated via reproduction of the ATLAS limits on Λ for the D5 and D9 effective operators with $m_\chi = 1$ GeV. Agreement is found within 12.5 and 7.4% respectively, where the ATLAS limits are consistently stronger, as shown in Table 8.7. The relative sizes of the discrepancies are expected given that only low-E_T^{miss} limits are available for the D5 operator while the high-E_T^{miss} signal region is used in this recast. Note

Table 8.7 The ATLAS mono-W/Z(had) E_T^{miss} signal region considered in this work and corresponding observed and expected model-independent upper limits on $\sigma \times \mathcal{A} \times \epsilon$ at 95% confidence level. (adapted from Ref. [41], using HistFitter)

Signal Region	E_T^{miss} threshold [GeV]	$\sigma \times \mathcal{A} \times \epsilon$ [fb] (obs)	$\sigma \times \mathcal{A} \times \epsilon$ [fb] (exp)
SR2	500	1.35	1.34

8.2 Recasting Mono-X Constraints

Table 8.8 Comparison of the 95% CL upper limits on Λ from this work and the 90% CL upper limits on Λ from the ATLAS mono-W/Z(had) analysis [41]. The limits are for the process $pp \to \chi\bar{\chi} + W/Z(\to jj)$

EFT operator	m_χ [GeV]	$\Lambda^{90\%\text{CL}}$ [GeV] (ATLAS)	$\Lambda^{95\%\text{CL}}$ [GeV] (this work)	Difference (%)
D9	1	2400	2221	7.4
D5	1	570	499	12.5

that a general discrepancy of a few percent is expected for both operators for the same reasons discussed above, and also because a cut-and-count approach is used while the ATLAS limits are extracted using a shape-fit. Furthermore, the ATLAS limits are quoted at 90% confidence level (CL) while these are calculated at 95% CL (Table 8.8).

8.3 Results and Discussion

8.3.1 Limits on the Couplings

The 95% CL upper limits on the coupling combination $\sqrt{g_q g_\chi}$ of the sV and sA models, and the tS model coupling $g_{q\chi}$, obtained from each of the mono-X channels, are presented in Figs. 8.3, 8.4, 8.5, 8.6, 8.7 and 8.8. They are evaluated as described in Sect. 8.2.2, including statistical and systematic uncertainties, and correspond to the best limits of each signal region tested.

In each plot, the limits are shown ranging from 0.01 to the upper perturbative limit, 4π. Where a limit was determined to be larger than this value, the limit is considered meaningless and the region is coloured grey. The white (hatched) regions coincide with those mass points which yield an initial (final) value of $\sqrt{g_q g_\chi}$ or $g_{q\chi}$ which fails to satisfy $\Gamma < M_{\text{med}}/2$. When $g_\chi/g_q = 0.2$, only the monojet channel produces sufficiently strong limits which survive this requirement, and so these are shown separately in Fig. 8.7.

The following model-independent trends are observed. For the sV model, strong limits exist in the regime where $M_{\text{med}} > 2m_\chi$ as the mediator can go on-shell and the cross section is enhanced. The sA model limits are generally similar to those for the sV model except in the region corresponding to $m_\chi \gtrsim \sqrt{4\pi} M_{\text{med}}/g_\chi^{\text{gen}}$ where g_χ^{gen} is the DM coupling used at the generator level. This region is removed in the sA model to avoid violating perturbative unitarity, which can lead to an unphysical enhancement of the cross section when m_χ is much larger than M_{med} [23, 24]. The upper limit on $\sqrt{g_q g_\chi}$ is relatively constant across different coupling ratios g_χ/g_q, as is expected when the coupling (and hence the width) has been demonstrated to have little effect on kinematic distributions (see Sect. 8.1.3), and using the assumptions of

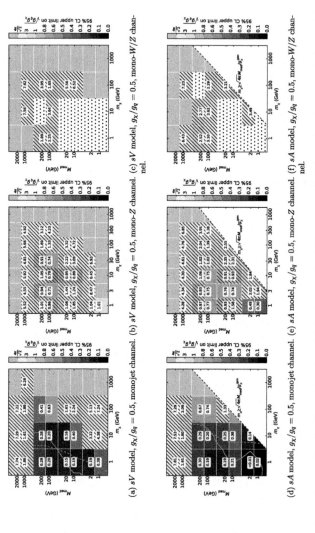

Fig. 8.3 Upper limits on the coupling for the s-channel models in the monojet (left), mono-Z(lep) (centre) and mono-W/Z(had) (right) channels, for $g_\chi/g_q = 0.5$. The grey region represents the phase space where no meaningful limit was obtained. The hatched region denotes where a limit is obtained that leads to a width greater than $M_{\text{med}}/2$, so the assumptions of the rescaling procedure begin to fail. The dotted region represents phase space where insufficient statistics of the signal simulation were available. Below and to the right of the dashed purple line, direct detection constraints become stronger than the mono-X constraint. The solid purple line shows which model points would be compatible with the relic density constraints: points above and to the left of the line the couplings leading to the correct relic density are naively ruled out, below and to the right of this line the relic density coupling is still allowed. See the text for further details

8.3 Results and Discussion

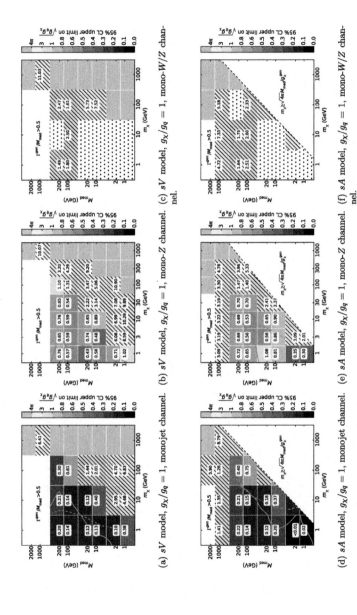

Fig. 8.4 Upper limits on the couplings for the s-channel models in the monojet (left), mono-Z(lep) (centre) and mono-W/Z(had) (right) channels, for $g_\chi/g_q = 1$. The same conventions as in Fig. 8.3 are used

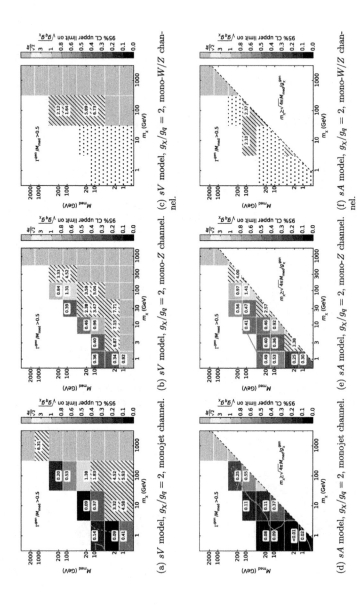

Fig. 8.5 Upper limits on the coupling for the *s*-channel models in the monojet (left), mono-Z(lep) (centre) and mono-W/Z(had) (right) channels, for $g_\chi/g_q = 2$. The same conventions as in Fig. 8.3 are used

8.3 Results and Discussion

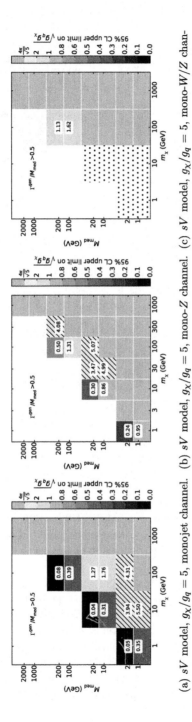

Fig. 8.6 Upper limits on the coupling for the sV model in the monojet (left), mono-Z(lep) (centre) and mono-W/Z(had) (right) channels, for $g_X/g_q = 5$. The same conventions as in Fig. 8.3 are used

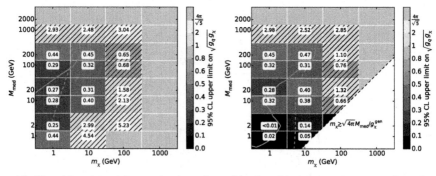

Fig. 8.7 Upper limits on the coupling for the s-channel models sV (left) and sA (right) in the monojet channel, for $g_\chi/g_q = 0.2$. The same conventions as in Fig. 8.3 are used

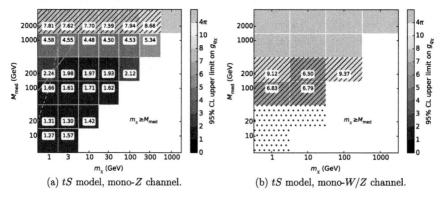

Fig. 8.8 Upper limits on the coupling $g_{q\chi}$ for the t-channel model in the mono-Z(lep) (left) and mono-W/Z(had) (right) channels. The same conventions as in Fig. 8.3 are used

Eq. 8.7. As the ratio of g_χ/g_q increases, points in the region $M_{\text{med}} > m_\chi$ disappear as the initial value, $g_q = 1$, leads to a failure of the width condition. However, by choosing a smaller initial value of g_q these points could in principle be recovered. The limits in this region would be expected to be similar to those seen in the $g_\chi/g_q = 0.2$ and 0.5 cases.

The constraints on the coupling strength are weaker when m_χ or M_{med} is large (>100 GeV) due to the reduction of the cross section. In this region, the constraints are expected to improve at higher centre-of-mass energies. For small DM masses with an off-shell mediator, the E_T^{miss} distribution is softer, therefore results in this region of phase space are limited by statistical uncertainties associated with the tail of the distribution. This region of phase space would benefit from dedicated optimisations of event selections aiming at the study of Simplified Models instead of EFTs.

In the following, detailed comments specific to each channel are discussed.

8.3.1.1 Monojet Channel

The upper limits on the coupling combination $\sqrt{g_q g_\chi}$ for the sV and sA models from the monojet channel are displayed in the left-hand column of Figs. 8.3, 8.4, 8.5 and 8.6, for $g_\chi/g_q = 0.5, 1, 2$ and 5 respectively (where the ratio of 5 is only shown for the sV model, due to a lack of meaningful results in the sA model). The $g_\chi/g_q = 0.2$ case is shown separately in Fig. 8.7, as meaningful limits are only obtained within this channel.

As expected, the monojet channel produces the strongest limits for both s-channel models, which are better than those from the next-best mono-Z(lep) channel by a factor of generally 1.5–10. The weakest limits are obtained for large m_χ or large M_{med}, where they might enter the regime where $\Gamma > M_{\text{med}}/2$ and the rescaling assumptions do not hold anymore. Although the acceptance is considerably higher when both m_χ and M_{med} are large, the cross section is sufficiently small so as to cancel any positive impact. Within the valid region ($m_\chi \in [1, 100]$ GeV and $M_{\text{med}} \in [1, 200]$ GeV), the limit on $\sqrt{g_q g_\chi}$ generally ranges from 0.1 to 0.7, with some on-shell masses reaching a limit of ~0.05 in the large g_χ/g_q case. In the large g_χ/g_q scenario, limits for $m_\chi = 1000$ GeV start to become valid: where $\sqrt{g_q g_\chi}$ remains constant but g_χ/g_q increases, the value of g_q is pushed downward and so the width, which is dominated by decays to SM particles, decreases with respect to m_χ.

The uncertainties on the limits for both s-channel models are dominated by contributions from the matching scale at acceptance-level, and generally range from about 5 to 46%.

8.3.1.2 Mono-Z(lep) Channel

The simplicity of the mono-Z(lep) channel relative to the monojet channel, and the ease of signal simulation at MADGRAPH level allowed to study a finer granularity of points in the mass phase space. The resulting limits on the sV and sA models are shown in the central column of Figs. 8.3, 8.4, 8.5 and 8.6. While the behaviour of the limits as g_χ/g_q is varied is similar to that seen in the monojet channel, the mono-Z(lep) limits are overall weaker.

The total relative uncertainties on $\sqrt{g_q g_\chi}$ for the s-channel models are generally within 10%, but can range up to 80% in a few cases where m_χ and M_{med} are very small. In general, they are split equally between statistical and systematic contributions.

As discussed above, one advantage of the mono-boson channels is that a study of the tS model is easily possible, due to the basically unchanged kinematic distributions when the couplings are varied. Since this is not the case for the tS model in the monojet channel, the strongest limits on tS in this study are obtained with the mono-Z(lep) analysis, and are shown in the left-hand side of Fig. 8.8. Note that, in comparison to the s-channel models, the limits have weakened by a factor of 10. This is the result of an order-of-magnitude weaker cross section and the inability of the mediator to go on-shell in this channel. Stronger limits are found for smaller m_χ and M_{med} masses, where larger cross sections compensate for lower acceptances. Overall, the statistical and systematic uncertainties contribute less than 10%.

8.3.1.3 Mono-W/Z(had) Channel

The limits on the couplings of the sV, sA and tS models, obtained within the mono-W/Z(had) channel, are shown in the right-hand column of Figs. 8.3, 8.4, 8.5, 8.6, 8.7 and 8.8. This channel was included for comparison with the leptonic mono-Z(lep) channel in particular, but a coarser selection of masses was chosen as the limits were initially found to be weaker. Additionally, two further assumptions were made. First, as the kinematic behaviour is reasonably independent of the couplings, a single acceptance was determined for each (m_χ, M_{med}) combination and applied to each value of g_χ/g_q. Second, complete systematic uncertainties were generated for a subset of masses and compared to those from the mono-Z(lep) channel. From this comparison the ratio between mono-Z(lep) and mono-W/Z(had) systematic uncertainties was determined and then applied to the other mass points. As a result, the limits obtained in this channel are not intended to be rigorously quantitative. Rather, they are used to indicate qualitatively how the channels compare.

The ATLAS mono-W/Z(had) analysis (and in particular the higher-E_T^{miss} signal region) was not optimised for a SiM interpretation, and much of the phase space produced insignificant numbers of events passing the event selection, with up to 200,000 events generated. Generally, the limits are weaker than those from the mono-Z(lep) channel by a factor between one and three, which is both consistent with the limits on the EFT models studied in the ATLAS analyses, and expected from the use of a cut-and-count interpretation in this study rather than a shape analysis like in the mono-W/Z(had) public results. In some points the limits become comparable with the mono-Z(lep) channel, suggesting that more statistics and an improved treatment of systematic uncertainties would bring these closer in line.

Overall, the uncertainties from this channel lie within 5 to 50% and are usually between 10 and 30%. Generally, both statistical and systematic uncertainties contribute in a similar manner. A few points are clearly limited by the generated statistics of the signal simulation, resulting in a statistical error of up to 90%. Points with high m_χ and low M_{med} tend to have larger systematic uncertainties.

8.3.2 Comparison with Relic Density Constraints

In Figs. 8.3, 8.4, 8.5, 8.6, 8.7 and 8.8 magenta lines indicate where the limit on the coupling corresponds to the coupling strength that would reproduce the correct DM relic density if DM was a thermal relic of the early universe. For points diagonally above and to the left of the solid purple line, the LHC constraints naively rule out the couplings leading to the correct relic density. Below and to the right of this line the relic density coupling is still allowed. For some scenarios the intercept does not pass

8.3 Results and Discussion

through a significant number of data points surviving the quality criteria outlined in previous sections. In these cases the line is not shown. The measured abundance is approximately related to the unknown self-annihilation cross section via:

$$\Omega_{\text{DM}} h^2 \simeq \frac{2 \times 2.4 \times 10^{-10} \,\text{GeV}^{-2}}{\langle \sigma v \rangle_{\text{ann}}}. \tag{8.10}$$

This is used with measurements of the DM abundance by Planck, $\Omega_{\text{DM}}^{\text{obs}} h^2 = 0.1199 \pm 0.0027$ [54], to find $\langle \sigma v \rangle_{\text{ann}} \simeq 4.0 \times 10^{-9} \,\text{GeV}^{-2}$ for thermal relic DM. The above relation (8.10) is only approximately accurate, and so the micrOMEGAs code [55] is used to determine the coupling strength leading to the correct relic density for each model.

The relic density couplings should by no means be regarded as strict constraints. If DM is not produced thermally or there is an unknown effect which modifies the evolution of the density with temperature, then Eq. 8.10 breaks down. Additionally, in the scenario where DM is assumed to be a thermal relic, the possibility of there being other annihilation channels and other new-physics particles contributing to the DM abundance is ignored, which, if taken into account, would also invalidate Eq. 8.10.

8.3.3 Comparison with Direct Detection Constraints

In Figs. 8.3, 8.4, 8.5, 8.6, 8.7 and 8.8 the intercept line where constraints from direct detection experiments are equivalent to mono-X constraints are shown as well. Below and to the right of the dashed purple line, direct detection constraints are stronger while above and to the left of this line, the LHC bounds are considered to be stronger. As with the relic density contours, the intercept is not shown where it does not pass through sufficient valid data points. The toolset from Ref. [56] has been used to convert the strongest available direct detection constraints at the time of the study, which come from the LUX 2013 dataset [57], onto constraints on the SiMs.

Compared to direct detection, the mono-X collider limits perform better for the sA model than for the sV model. This is because the axial-vector coupling leads to a suppressed scattering rate in direct detection experiments while collider searches are relatively insensitive to the difference between the vector and axial-vector couplings. In the non-relativistic limit, the tS model leads to a mix of both suppressed and unsuppressed operators.

The direct detection constraints assume that the DM candidate under consideration contributes 100% of the local DM density, while the mono-X constraints make no assumptions about either the local DM density or overall abundance. In this sense the mono-X limits remain useful even in those regions of phase space where they are not as strong as those from direct detection.

8.4 Conclusions

Simplified Models of DM production allow for an improved way of interpreting LHC searches for DM in a rather general and model-independent way, avoiding validity issues that affect results obtained within effective field theory models. Constraints from ATLAS Run I mono-X searches in the monojet, mono-$Z(\to$ leptons), and mono-$W/Z(\to$ hadrons) channels have been re-interpreted in terms of three Simplified Models and their parameters. Rather than setting limits in the $M_{\text{med}} - m_\chi$ plane for a fixed value of the coupling strength, the coupling strength is constrained as a function of both M_{med} and m_χ. While this approach necessitates the introduction of some approximations, it also allows for a thorough examination of the interplay between the DM production cross section and the free parameters of the models. The region of parameter space where both the DM and mediator masses span $\mathcal{O}(\text{GeV})$ to $\mathcal{O}(\text{TeV})$ has been explored, and coupling ratios g_χ/g_q of 0.2, 0.5, 1, 2 and 5 were considered. This study significantly extended the Simplified-Model interpretations that were performed within the experimental analyses.

As expected, the monojet channel is found to yield the strongest limits on vector and axial-vector couplings to a vector mediator exchanged in the s-channel. The monojet channel is also found to perform well for smaller values of g_χ. The limits obtained in the mono-Z(lep) channel, in comparison, are generally weaker, while the mono-W/Z(had) results are weaker again. This is partly due to the conservative estimation of the systematic uncertainties and partly due to limited statistics resulting from the E_T^{miss} selection cut often not being appropriate for the regarded Simplified Model. The width effects associated with the t-channel exchange of an $SU(2)$ doublet scalar mediator are observed to vanish in both the mono-Z(lep) and mono-W/Z(had) channels, greatly simplifying the analysis and confirming these as straightforward and competitive channels for future collider DM detection. Where the axial-vector model is not excluded by perturbative unitarity requirements, the coupling limits are found to be similar to those of the vector model within each analysis channel. Weaker limits are found for the t-channel model, a result of cross section suppression that is not present in the s-channel models.

It is important to note that the mono-X searches are complementary to direct searches for the mediator, e.g. via dijet resonances [58–61]. These have been used to study SiMs in, for example, Refs. [17, 24, 30]. Dijet studies search for the signature of a mediator decay into Standard Model particles, generally assuming a narrow resonance. These constraints can be stronger than mono-X constraints, particularly when the width is small and when the mediator coupling to quarks is stronger than the one to DM. Mono-X searches however are advantageous for larger values of g_χ/g_q, larger widths and smaller mediator masses. With the use of Simplified Models, this aspect was significantly strengthened and within the experimental collaborations much more attention is given to the possible interplay between mono-X and mediator searches, manifesting itself e.g. in a joint summary plot, as was discussed in Sect. 7.8.

Furthermore, this study revealed that selection optimisations in view of Simplified Models would be especially beneficial for the low-E_T^{miss} regime of smaller M_{med} and

8.4 Conclusions

m_{DM}: while EFT models strongly suggested a focus on the high-E_T^{miss} region, Simplified Models motivate to also look at low-E_T^{miss} signatures. This is experimentally often more challenging and probably will require to include shape-fit techniques into the mono-X analyses in the future.

Finally, the limits are compared to constraints from relic density and direct detection. Although each search direction needs to make different assumptions, this demonstrates the complementarity of the searches and the importance of Simplified Models as a tool for the interpretation of collider DM searches.

References

1. J. Goodman, M. Ibe, A. Rajaraman, W. Shepherd, T.M.P. Tait, H.-B. Yu, Constraints on Dark Matter from Colliders. Phys. Rev. D **82**, 116010 (2010). https://doi.org/10.1103/PhysRevD.82.116010, arXiv:1008.1783 [hep-ph]
2. Y. Bai, P.J. Fox, R. Harnik, The Tevatron at the Frontier of Dark Matter Direct Detection. JHEP **12**, 048 (2010). https://doi.org/10.1007/JHEP12(2010)048, arXiv:1005.3797 [hep-ph]
3. P.J. Fox, R. Harnik, J. Kopp, Y. Tsai, LEP Shines Light on Dark Matter. Phys. Rev. D **84**, 014028 (2011). https://doi.org/10.1103/PhysRevD.84.014028, arXiv:1103.0240 [hep-ph]
4. M. L. Graesser, I. M. Shoemaker, L. Vecchi, A Dark Force for Baryons, arXiv:1107.2666 [hep-ph]
5. H. An, F. Gao, Fitting CoGeNT Modulation with an Inelastic, Isospin-Violating Z' Model, arXiv:1108.3943 [hep-ph]
6. O. Buchmueller, M.J. Dolan, C. McCabe, Beyond effective field theory for dark matter searches at the LHC. JHEP **01**, 025 (2014). https://doi.org/10.1007/JHEP01(2014)025, arXiv:1308.6799 [hep-ph]
7. G. Busoni, A. De Simone, E. Morgante, A. Riotto, On the validity of the effective field theory for dark matter searches at the LHC. Phys. Lett. B **728**, 412–421 (2014). https://doi.org/10.1016/j.physletb.2013.11.069, arXiv:1307.2253 [hep-ph]
8. G. Busoni, A. De Simone, J. Gramling, E. Morgante, A. Riotto, On the validity of the effective field theory for dark matter searches at the LHC, part II: complete analysis for the s-channel. JCAP **1406**, 060 (2014). https://doi.org/10.1088/1475-7516/2014/06/060, arXiv:1402.1275 [hep-ph]
9. G. Busoni, A. De Simone, T. Jacques, E. Morgante, A. Riotto, On the validity of the effective field theory for dark matter searches at the LHC part III: analysis for the t-channel. JCAP **1409**, 022 (2014). https://doi.org/10.1088/1475-7516/2014/09/022, arXiv:1405.3101 [hep-ph]
10. A. DiFranzo, K.I. Nagao, A. Rajaraman, T.M.P. Tait, Simplified models for dark matter interacting with quarks. JHEP **11**, 014 (2013). https://doi.org/10.1007/JHEP11(2013)014,162, arXiv:1308.2679 [hep-ph]
11. M.R. Buckley, D. Feld, D. Goncalves, Scalar simplified models for dark matter. Phys. Rev. D **91**, 015017 (2015). https://doi.org/10.1103/PhysRevD.91.015017, arXiv:1410.6497 [hep-ph]
12. A.J. Brennan, M.F. McDonald, J. Gramling, T.D. Jacques, Collide and conquer: constraints on simplified dark matter models using mono-X collider searches. JHEP **05**, 112 (2016). https://doi.org/10.1007/JHEP05(2016)112, arXiv:1603.01366 [hep-ph]
13. R.M. Godbole, G. Mendiratta, T.M.P. Tait, A simplified model for dark matter interacting primarily with gluons, *JHEP***08**(2015) 064. https://doi.org/10.1007/JHEP08(2015)064, arXiv:1506.01408 [hep-ph]
14. O. Buchmueller, M.J. Dolan, S.A. Malik, C. McCabe, Characterising dark matter searches at colliders and direct detection experiments: vector mediators. JHEP **01**, 037 (2015). https://doi.org/10.1007/JHEP01(2015)037, arXiv:1407.8257 [hep-ph]

15. M. Blennow, J. Herrero-Garcia, T. Schwetz, S. Vogl, Halo-independent tests of dark matter direct detection signals: local DM density, LHC, and thermal freeze-out. JCAP **1508**(08), 039 (2015). https://doi.org/10.1088/1475-7516/2015/08/039, arXiv:1505.05710 [hep-ph]
16. A. Alves, A. Berlin, S. Profumo, F.S. Queiroz, Dark matter complementarity and the Z' Portal. Phys. Rev. D **92**(8), 083004 (2015). https://doi.org/10.1103/PhysRevD.92.083004, arXiv:1501.03490 [hep-ph]
17. H. An, X. Ji, L.-T. Wang, Light Dark Matter and Z' Dark force at colliders. JHEP **07**, 182 (2012). https://doi.org/10.1007/JHEP07(2012)182, arXiv:1202.2894 [hep-ph]
18. H. An, R. Huo, L.-T. Wang, Searching for low mass dark portal at the LHC. Phys. Dark Univ. **2**, 50–57 (2013). https://doi.org/10.1016/j.dark.2013.03.002, arXiv:1212.2221 [hep-ph]
19. M.T. Frandsen, F. Kahlhoefer, A. Preston, S. Sarkar, and K. Schmidt-Hoberg, LHC and Tevatron bounds on the dark matter direct detection cross-section for vector mediators, *JHEP***07**(2012) 123. https://doi.org/10.1007/JHEP07(2012)123, arXiv:1204.3839 [hep-ph]
20. G. Arcadi, Y. Mambrini, M.H.G. Tytgat, and B. Zaldivar, Invisible Z' and dark matter: LHC vs LUX constraints," *JHEP***03**(2014) 134. https://doi.org/10.1007/JHEP03(2014)134, arXiv:1401.0221 [hep-ph]
21. P. Harris, V.V. Khoze, M. Spannowsky, C. Williams, Constraining dark sectors at colliders: beyond the effective theory approach. Phys. Rev. D **91**, 055009 (2015). https://doi.org/10.1103/PhysRevD.91.055009. arXiv:1411.0535 [hep-ph]
22. T. Jacques, K. Nordström, Mapping monojet constraints onto simplified dark matter models. JHEP **06**, 142 (2015). https://doi.org/10.1007/JHEP06(2015)142, arXiv:1502.05721 [hep-ph]
23. N.F. Bell, Y. Cai, R.K. Leane, Mono-W dark matter signals at the LHC: simplified model analysis. JCAP **1601**(01), 051 (2016). https://doi.org/10.1088/1475-7516/2016/01/051, arXiv:1512.00476 [hep-ph]
24. M. Chala, F. Kahlhoefer, M. McCullough, G. Nardini, K. Schmidt-Hoberg, Constraining dark sectors with monojets and dijets. JHEP **07**(2015) 089. https://doi.org/10.1007/JHEP07(2015)089, arXiv:1503.05916 [hep-ph]
25. F. Kahlhoefer, K. Schmidt-Hoberg, T. Schwetz, S. Vogl, Implications of unitarity and gauge invariance for simplified dark matter models. JHEP **02**(2016) 016. https://doi.org/10.1007/JHEP02(2016)016, arXiv:1510.02110 [hep-ph]
26. G. Jungman, M. Kamionkowski, K. Griest, Supersymmetric dark matter. Phys. Rept.**267** (1996) 195–373. https://doi.org/10.1016/0370-1573(95)00058-5, arXiv:hep-ph/9506380 [hep-ph]
27. Y. Baim, J. Berger, Fermion portal dark matter. JHEP **11**, 171 (2013). https://doi.org/10.1007/JHEP11(2013)171, arXiv:1308.0612 [hep-ph]
28. H. An, L.-T. Wang, H. Zhang, Dark matter with t-channel mediator: a simple step beyond contact interaction. Phys. Rev. D **89**(11), 115014 (2014). https://doi.org/10.1103/PhysRevD.89.115014, arXiv:1308.0592 [hep-ph]
29. S. Chang, R. Edezhath, J. Hutchinson, M. Luty, Effective WIMPs. Phys. Rev. D **89**(1), 015011 (2014). https://doi.org/10.1103/PhysRevD.89.015011. arXiv:1307.8120 [hep-ph]
30. M. Papucci, A. Vichi, K.M. Zurek, Monojet versus the rest of the world I: t-channel models. JHEP **11**, 024 (2014). https://doi.org/10.1007/JHEP11(2014)024, arXiv:1402.2285 [hep-ph]
31. M. Garny, A. Ibarra, S. Vogl, Signatures of Majorana dark matter with t-channel mediators. Int. J. Mod. Phys. D **24**(07), 1530019 (2015). https://doi.org/10.1142/S0218271815300190, arXiv:1503.01500 [hep-ph]
32. M. Garny, A. Ibarra, S. Rydbeck, S. Vogl, Majorana dark matter with a coloured mediator: collider vs direct and indirect searches. JHEP **06**, 169 (2014). https://doi.org/10.1007/JHEP06(2014)169, arXiv:1403.4634 [hep-ph]
33. N.F. Bell, J.B. Dent, A.J. Galea, T.D. Jacques, L.M. Krauss, T.J. Weiler, W/Z bremsstrahlung as the dominant annihilation channel for dark matter, revisited. Phys. Lett. B **706**, 6–12 (2011). https://doi.org/10.1016/j.physletb.2011.10.057, arXiv:1104.3823 [hep-ph]
34. ATLAS Collaboration, Search for dark matter produced in association with a hadronically decaying vector boson in pp collisions at $\sqrt{s} = 13$ TeV with the ATLAS detector at the LHC, Techcnical Report ATLAS-CONF-2015-080, CERN, Geneva, Dec, 2015, http://cds.cern.ch/record/2114852

35. P.J. Fox, R. Harnik, J. Kopp, Y. Tsai, Missing energy signatures of dark matter at the LHC. Phys. Rev. D **85**, 056011 (2012). https://doi.org/10.1103/PhysRevD.85.056011, arXiv:1109.4398 [hep-ph]
36. P.J. Fox, R. Harnik, R. Primulando, C.-T. Yu, Taking a razor to dark matter parameter space at the LHC. Phys. Rev. D **86**, 015010 (2012). https://doi.org/10.1103/PhysRevD.86.015010, arXiv:1203.1662 [hep-ph]
37. ATLAS Collaboration, G. Aad et al., Search for new phenomena in final states with an energetic jet and large missing transverse momentum in pp collisions at $\sqrt{s}$ =8 TeV with the ATLAS detector. Eur. Phys. J. **C75**(7), 299 (2015). https://doi.org/10.1140/epjc/s10052-015-3517-3, https://doi.org/10.1140/epjc/s10052-015-3639-7, arXiv:1502.01518 [hep-ex]
38. D. Abercrombie et al., Dark matter benchmark models for early lhc run-2 searches: report of the ATLAS/CMS dark matter forum, arXiv:1507.00966 [hep-ex]
39. DELPHES 3 Collaboration, J. de Favereau, C. Delaere, P. Demin, A. Giammanco, V. Lemaître, A. Mertens, and M. Selvaggi, DELPHES 3, a modular framework for fast simulation of a generic collider experiment, JHEP **02**(2014) 057. https://doi.org/10.1007/JHEP02(2014)057, arXiv:1307.6346 [hep-ex]
40. ATLAS Collaboration, G. Aad et al., Search for dark matter in events with a Z boson and missing transverse momentum in pp collisions at $\sqrt{s}$=8 TeV with the ATLAS detector, Phys. Rev. **D90**(1), 012004 (2014). https://doi.org/10.1103/PhysRevD.90.012004, arXiv:1404.0051 [hep-ex]
41. ATLAS Collaboration, G. Aad et al., Search for dark matter in events with a hadronically decaying W or Z boson and missing transverse momentum in pp collisions at $\sqrt{s} = 8$ TeV with the ATLAS detector. Phys. Rev. Lett. **112**(4), 041802 (2014). https://doi.org/10.1103/PhysRevLett.112.041802, arXiv:1309.4017 [hep-ex]
42. J. Alwall, R. Frederix, S. Frixione, V. Hirschi, F. Maltoni, O. Mattelaer, H.S. Shao, T. Stelzer, P. Torrielli, M. Zaro, The automated computation of tree-level and next-to-leading order differential cross sections, and their matching to parton shower simulations. JHEP **1407**, 079 (2014). https://doi.org/10.1007/JHEP07(2014)079, arXiv:1405.0301 [hep-ph]
43. A.D. Martin, W.J. Stirling, R.S. Thorne, G. Watt, Parton distributions for the LHC. Eur. Phys. J. C **63**, 189–285 (2009). https://doi.org/10.1140/epjc/s10052-009-1072-5, arXiv:0901.0002 [hep-ph]
44. T. Sjöstrand, S. Ask, J.R. Christiansen, R. Corke, N. Desai, P. Ilten, S. Mrenna, S. Prestel, C.O. Rasmussen, P.Z. Skands, An Introduction to PYTHIA 8.2. Comput. Phys. Commun. **191**, 159–177 (2015). https://doi.org/10.1016/j.cpc.2015.01.024, arXiv:1410.3012 [hep-ph]
45. ATLAS Collaboration, Summary of ATLAS Pythia 8 tunes, http://cds.cern.ch/record/1474107
46. M. Cacciari, G.P. Salam, G. Soyez, FastJet User Manual. Eur. Phys. J. C **72**, 1896 (2012). https://doi.org/10.1140/epjc/s10052-012-1896-2, arXiv:1111.6097 [hep-ph]
47. ATLAS Collaboration, G. Aad et al., Performance of jet substructure techniques for large-R jets in proton-proton collisions at $\sqrt{s} = 7$ TeV using the ATLAS detector, JHEP **09**, 076 (2013). https://doi.org/10.1007/JHEP09(2013)076, arXiv:1306.4945 [hep-ex]
48. M.L. Mangano, M. Moretti, F. Piccinini, M. Treccani, Matching matrix elements and shower evolution for top-quark production in hadronic collisions. JHEP **01**, 013 (2007), arXiv:hep-ph/0611129 [hep-ph]
49. S. Schramm, P. Savard, Searching for dark matter with the ATLAS detector in events with an energetic jet and large missing transverse momentum. Ph.D. Thesis, Toronto U., Mar 2015, http://cds.cern.ch/record/2014029. Presented 26 Feb 2015
50. P. Skands, S. Carrazza, J. Rojo, Tuning PYTHIA 8.1: the Monash 2013 tune, Eur. Phys. J. **C74**(8), 3024 (2014). https://doi.org/10.1140/epjc/s10052-014-3024-y, arXiv:1404.5630 [hep-ph]
51. A. Wald, Sequential tests of statistical hypotheses. Ann. Math. Stat. **16**(2), 117–186 (1945). https://doi.org/10.1214/aoms/1177731118
52. M. Baak, G.J. Besjes, D. Côte, A. Koutsman, J. Lorenz, D. Short, HistFitter software framework for statistical data analysis. Eur. Phys. J. C **75**, 153 (2015). https://doi.org/10.1140/epjc/s10052-015-3327-7, arXiv:1410.1280 [hep-ex]

53. N.F. Bell, Y. Cai, J.B. Dent, R.K. Leane, T.J. Weiler, Dark matter at the LHC: effective field theories and gauge invariance. Phys. Rev. D **92**(5), 053008 (2015). https://doi.org/10.1103/PhysRevD.92.053008. arXiv:1503.07874 [hep-ph]
54. P.A.R. Planck, Collaboration, Ade et al., Planck 2013 results. XVI. Cosmological parameters. Astron. Astrophys. **571**, A16 (2014). https://doi.org/10.1051/0004-6361/201321591. arXiv:1303.5076 [astro-ph.CO]
55. G. Bélanger, F. Boudjema, A. Pukhov, A. Semenov, micrOMEGAs4.1: two dark matter candidates. Comput. Phys. Commun. **192**, 322–329 (2015). https://doi.org/10.1016/j.cpc.2015.03.003, arXiv:1407.6129 [hep-ph]
56. M. Cirelli, E. Del Nobile, P. Panci, Tools for model-independent bounds in direct dark matter searches. JCAP **1310**, 019 (2013). https://doi.org/10.1088/1475-7516/2013/10/019. arXiv:1307.5955 [hep-ph]
57. L.U.X. Collaboration, D.S. Akerib et al., First results from the LUX dark matter experiment at the Sanford Underground Research Facility. Phys. Rev. Lett. **112**, 091303 (2014). https://doi.org/10.1103/PhysRevLett.112.091303, arXiv:1310.8214 [astro-ph.CO]
58. C.M.S. Collaboration, S. Chatrchyan et al., Search for narrow resonances using the dijet mass spectrum in pp collisions at $\sqrt{s}$=8 TeV. Phys. Rev. D **87**(11), 114015 (2013). https://doi.org/10.1103/PhysRevD.87.114015, arXiv:1302.4794 [hep-ex]
59. ATLAS Collaboration, G. Aad et al., Search for new phenomena in the dijet mass distribution using $p - p$ collision data at $\sqrt{s} = 8$ TeV with the ATLAS detector, Phys. Rev. D **91**(5), 052007 (2015). https://doi.org/10.1103/PhysRevD.91.052007, arXiv:1407.1376 [hep-ex]
60. C.D.F. Collaboration, T. Aaltonen et al., Search for new particles decaying into dijets in proton-antiproton collisions at $\sqrt{s} = 1.96$ TeV. Phys. Rev. D **79**, 112002 (2009). https://doi.org/10.1103/PhysRevD.79.112002, arXiv:0812.4036 [hep-ex]
61. C.M.S. Collaboration, V. Khachatryan et al., Search for resonances and quantum black holes using dijet mass spectra in proton-proton collisions at $\sqrt{s} = 8$ TeV. Phys. Rev. D **91**(5), 052009 (2015). https://doi.org/10.1103/PhysRevD.91.052009, arXiv:1501.04198 [hep-ex]

Chapter 9
Search for New Physics in Events with Missing Energy and Top Quarks

The supersymmetric partner of the top quark, the *stop*, might be significantly lighter than the other squarks, as was motivated in Chap. 4. Hence, the search for stop production at the LHC presents a well-motivated approach to look for SUSY at the LHC. The stop can decay in different ways, depending on the SUSY particle mass spectrum, in particular on the masses of the stops $\tilde{t}_1$ and $\tilde{t}_2$, the charginos $\tilde{\chi}^\pm$ and the lightest neutralino $\tilde{\chi}_1^0$, and other model parameters. The analysis presented in the following considers two possible stop decay scenarios, illustrated in Fig. 9.1. Both scenarios assume R-parity conservation, hence the stops are produced in pairs. In the first scenario, the stop decays into a top quark and the lightest neutralino: $\tilde{t}_1 \to t + \tilde{\chi}_1^0$ (tN), the second scenario assumes the decay into a b-quark and the lightest chargino, where the latter decays further into a W boson and the lightest neutralino: $\tilde{t}_1 \to b + \tilde{\chi}_1^\pm$ (bC). Furthermore, a *mixed* decay scenario where both decay channels are allowed with various branching ratio ($\mathcal{BR}$) assumptions is considered in the interpretation.

The pair production of Dark Matter (DM) particles in association with top quarks, illustrated in Fig. 9.2, is motivated by the assumption of (pseudo-)scalar mediators with Yukawa-like quark-mass dependent couplings to the Standard Model sector [1–3]. Such a signal presents the same final state as the tN stop decay scenario described above: in both cases a $t\bar{t}$ pair is produced together with undetectable particles–the neutralinos or the DM particles–that lead to missing transverse energy (E_T^{miss}). To profit from this similarity, the search for Dark Matter in association with top quarks is conducted together with the search for top squarks. The optimisation, design, background estimation and results of the DM signal regions will be the focus of the presentation of the analysis in this chapter.

The scenarios described above can lead to fully-hadronic, one-leptonic or dileptonic final states, depending on the decay mode of the W bosons that appear in the decay chain. The analysis of the one-lepton channel is presented in the following, hence the W boson from one of the top quarks is considered to decay to an electron or muon (either directly or via a tau lepton) and the W boson from the other top quark

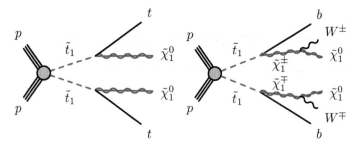

Fig. 9.1 Schematic description of the direct pair production of $\tilde{t}_1$ particles and their decays considered in this analysis, which are referred to as $\tilde{t}_1 \to t + \tilde{\chi}_1^0$ (left) and $\tilde{t}_1 \to b + \tilde{\chi}_1^\pm$ (right)

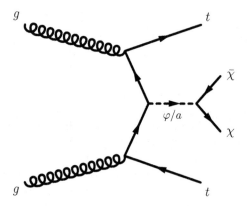

Fig. 9.2 Illustration of the production of Dark Matter particles in association with top quarks

decays hadronically. This channel has the advantage of being cleaner and having less background than the fully-hadronic channel while having a branching ratio almost as high. Analyses of the same dataset have been published as well in the zero- and two-lepton channels [4, 5], but are not further discussed here.

This analysis uses ATLAS data collected in proton-proton (pp) collisions at a center-of-mass energy of $\sqrt{s} = 13$ TeV, corresponding to an integrated luminosity of 13.2 fb^{-1}. While previous searches for DM in association with heavy quarks were based on Effective Field Theory (EFT) models [6, 7], this analysis considers a model with a (pseudo-) scalar mediator [2, 3]. It presents the first search for Dark Matter associated production with tops considering such Simplified Models. Furthermore, this analysis extends the previous search for top squarks that was performed with 3.2 fb^{-1} of 2015 data [8].

For this analysis, I performed the sensitivity studies and the cut optimisation and developed the search strategy for signals of Dark Matter with heavy flavour. I contributed to the development and validation of the software framework for this analysis. Emphasis is also put on a study of the behaviour of relevant triggers which I conducted. I was in charge of the production of the pre-selected data format that was used analysis-wide, the analysis-wide estimation of backgrounds as well as the statistical interpretation of the results. Furthermore, I conducted the limit-setting for Dark Matter Simplified Models.

After the general analysis strategy is presented in Sect. 9.1, Sect. 9.2 details the considered dataset and simulated samples, including a discussion of the trigger strategy and performance. The event selection and its optimisation is outlined in Sect. 9.3 and the main discriminating variables are introduced in Sect. 9.4. The selection optimisation for Dark Matter signals and the final signal region (SR) definitions are summarised in Sect. 9.5. Subsequently, Sect. 9.6 elucidates the background estimation and presents the definitions of the control and validation regions. Systematic uncertainties are discussed in Sect. 9.7. In Sect. 9.8, the measured event yields in data are presented together with the background expectation. The results are interpreted in terms of limits on top squark and Dark Matter production in Sect. 9.9. Finally, concluding remarks are given in Sect. 9.10.

9.1 Analysis Strategy

The analysis profits from dedicated, optimised selections for each signal scenario, which define the different SRs. The main backgrounds for this analysis are given by $t\bar{t}$ and W+jets events. They are efficiently reduced by cuts on discriminating variables, specifically designed for the targeted scenarios. In the final selection, also backgrounds from $t\bar{t} + Z(\to \nu\bar{\nu})$ and single top processes contribute significantly. The analysis estimates the expected background yields using Monte Carlo simulated samples which are normalised in background-enriched control regions. The extrapolation of this normalisation from control to signal region is verified in validation regions, kinematically situated between control and signal region. Via an extrapolation in the transverse mass, the $t\bar{t}$ and W+jets control and validation regions are defined for each SR. The single top control region selection relies in addition on a higher b-jet multiplicity and their angular distribution. In order to constrain the $t\bar{t}Z$ contribution, a $t\bar{t}\gamma$ control sample is defined. In a simultaneous fit to all control regions, the background normalisation and the resulting background expectation in the SRs is determined.

9.2 Dataset

The LHC pp collision data used in this analysis was recorded during 2015 and 2016 at a centre-of-mass energy of $\sqrt{s} = 13$ TeV, with a mean number of additional pp interactions per bunch crossing (*pile-up*) of 14 (in 2015) and 23.5 (in 2016).

Collisions that fulfil basic quality requirements and were recorded during stable beam and detector conditions amount to an integrated luminosity of $13.2\,\text{fb}^{-1}$. The uncertainty on this value is determined following Refs. [9, 10] and amounts to 2.1% for the 2015 dataset of 3.2 fb^{-1} and 3.7% for 2016 (10 fb^{-1}).

Table 9.1 Overview of the simulated samples

Process	ME generator	ME PDF	PS and hadronization	UE tune	Cross section order
$t\bar{t}$	POWHEG-BOX v2	CT10	PYTHIA 6	P2012	NNLO + NNLL
Single top	POWHEG-BOX v1/v2	CT10	PYTHIA 6	P2012	NNLO + NNLL
W/Z+jets	SHERPA 2.2	NNPDF3.0 NNLO	SHERPA	Default	NNLO
Diboson	SHERPA 2.1.1	CT10	SHERPA	Default	NLO
$t\bar{t} + W/Z$	MG5_aMC 2.2.2	NNPDF2.3	PYTHIA 8	A14	NLO
$t\bar{t} + \gamma$	MG5_aMC 2.2.3	CTEQ6L1	PYTHIA 8	A14	NLO
$W + \gamma$	SHERPA 2.1.1	CT10	SHERPA	Default	LO
SUSY signal	MG5_aMC 2.2.2	NNPDF2.3	PYTHIA 8	A14	NLO + NLL
DM signal	MG5_aMC 2.2.2	NNPDF3.0 LO	PYTHIA 8	A14	LO

9.2.1 Monte Carlo Simulations

Monte Carlo (MC) simulated samples are used to develop selections that discriminate well between signal and background by studying their different kinematic behaviours. The MC samples are also used to study detector acceptance and reconstruction efficiencies for signal and background processes and are needed in the background estimation.

Background Samples Different matrix element (ME) simulations are interfaced with implementations of parton shower and hadronisation in order to model the Standard Model processes which represent a background to this analysis. The details are summarised in Table 9.1. All samples are normalised to the cross section calculated up to the highest order in α_S available. The cross sections of the $t\bar{t}$, W+jets, and single top processes are only used for cross-checks and optimisation studies, while for the final results these processes are scaled by a normalisation that is determined from comparison with data.

The $t\bar{t}$ samples, as well as the single top samples, are generated with POWHEG [11], interfaced to PYTHIA 6 [12] for parton showering and hadronisation. They consider the re-summation of soft gluon emission to next-to-next-to-leading-logarithmic (NNLL) and next-to-leading-logarithmic (NLL) accuracy, respectively. In order to enhance the statistics in the tail of the E_T^{miss} distribution, inclusive samples are combined with samples applying a generator-level E_T^{miss} cut of 200 GeV.

The W+jets, Z+jets and diboson samples are generated with SHERPA [15] at leading order. They employ a simplification of the scale setting procedure in the multi-parton matrix elements. This allows for faster event generation. The jet multiplicity

9.2 Dataset

distribution is then reweighted at event level, where the correction factor is determined from an event generation using the strict scale prescription.

MADGRAPH [14] interfaced to PYTHIA 8 [31] is used to simulate processes of $t\bar{t}Z$ and $t\bar{t}W$.

Since a $t\bar{t}\gamma$ control region is used in the following to constrain the $t\bar{t}Z$ background, it is necessary that these processes are simulated as similarly as possible. The effect of the different choice of PDF set, factorisation and renormalisation scales and number of additional partons derived from the matrix element is accounted for by correcting the $t\bar{t}\gamma$ cross section by 4%.[1] The same NLO QCD k-factor is then applied to the $t\bar{t}\gamma$ process as used for the $t\bar{t}Z$ process. This choice is motivated by the similarity of QCD calculations for the two processes as well as empirical studies of the ratio of k-factors computed as a function of the boson p_T. The possible overlap between the $t\bar{t}\gamma$ sample and the $t\bar{t}$ sample, due to photons from final state radiation is considered in the following way: events containing photons that do not originate from hadronic decays or interactions with detector material, are removed from the $t\bar{t}$ sample if their transverse momentum exceeds 80 GeV and are removed from the $t\bar{t}\gamma$ sample if the photon p_T is below 80 GeV.[2]

SUSY Signal Samples The production of unpolarised stop pairs is generated at leading order (LO) using MADGRAPH [14] for the ME calculation.[3] The result is interfaced with PYTHIA 8 [31] which also calculates the stop decays. Different stop decay and mass configurations are considered.

First, a $\tilde{t}_1 \to t + \tilde{\chi}_1^0$ decay with a $\mathcal{BR}$ of 100% is assumed. Relevant samples are generated in a grid across the plane of $\tilde{t}_1$ and $\tilde{\chi}_1^0$ masses[4] with a spacing of 50 GeV. The granularity is increased towards the "diagonal" region where $m_{\tilde{t}_1}$ approaches $m_t + m_{\tilde{\chi}_1^0}$.

Second, a $\tilde{t}_1 \to b + \tilde{\chi}_1^\pm \to bW^{(*)}\tilde{\chi}_1^0$ decay is considered with a $\mathcal{BR}$ of 100%. Here, the parameter space is three-dimensional, spanned by the $\tilde{t}_1$, $\tilde{\chi}_1^\pm$, and $\tilde{\chi}_1^0$ masses. It is probed under two assumptions on the mass relations of the sparticles in the decay: the chargino mass is either set to twice the mass of the lightest neutralino ($m_{\tilde{\chi}_1^\pm} = 2m_{\tilde{\chi}_1^0}$),[5] or the chargino is taken to be slightly lighter than the stop, $m_{\tilde{\chi}_1^\pm} = m_{\tilde{t}_1} - 10$ GeV.

---

[1] The $t\bar{t}\gamma$ simulation fixes the factorisation and renormalisation scale to $2m_{top}$ and does not assume extra partons in the ME calculation. The $t\bar{t}Z$ sample on the other hand is simulated with up to two additional partons and using $\sum m_T$ as factorisation and renormalisation scale. To account for photon radiation from the top decay products, the top decay is described by MADGRAPH for $t\bar{t}\gamma$. The effect is of about 10% for $p_T^\gamma \sim 145$ GeV [16].

[2] The value of 80 GeV is motivated by a generator-level filter on the photon p_T applied in the generation of the $t\bar{t}\gamma$ sample.

[3] The re-summation of soft gluon emission to next-to-leading-logarithmic (NLL) accuracy is considered in the cross section.

[4] The $\tilde{\chi}_1^0$ is taken to be a pure bino.

[5] This choice is motivated from models that assume universal gaugino masses at the unification scale. Calculating their evolution using the renormalisation group equations leads the condition $m_{\tilde{\chi}_1^\pm} = 2m_{\tilde{\chi}_1^0}$ [17].

DM Signal Samples Samples of DM pair production are generated with MADGRAPH (MG5_aMC) using a Simplified-Model implementation [2, 3] corresponding to the diagram in Fig. 9.2. A scalar or pseudo-scalar mediator is assumed to profit from the enhancement of the coupling to heavy quarks. The masses of the mediator and DM particles, the couplings of the mediator to Dark Matter and Standard Model quarks, and the width of the mediator represent the free parameters of the model. The couplings of the mediator to the DM particles (g_χ) is taken to be equal to the mediator coupling to the quarks (g_q). For the common coupling defined as $g = \sqrt{g_q g_\chi}$ values between 0.1 and 3.5 are assumed. The minimal width of the mediator is calculated and then assumed for each model point [2]. The signal grid is defined in the plane of DM and mediator mass.

The detector response of ATLAS is simulated and applied to all generated samples. All background samples, except for $t\bar{t} + \gamma$, are processed with a full GEANT 4-based simulation [18], where for the signals and the $t\bar{t} + \gamma$ sample a fast simulation [19] that exploits a parameterisation of calorimeter showers, is used.

In all samples, the hard scattering is overlaid with varying numbers of simulated minimum-bias interactions generated with PYTHIA 8 to model the effect of pile-up, i.e. simulating multiple pp interactions that might occur in the same or nearby bunch crossings. The average number of pile-up interactions is reweighted to the distribution measured in data. To account for differences in the object reconstruction and identification efficiencies between data and MC, the simulated samples are reweighted accordingly [20, 24].

9.2.2 Trigger Selection

Events used for this search are accepted for recording by a E_T^{miss}-based trigger logic, which is purely based on calorimetric information. The trigger used for the 2015 dataset is xe80_tc_lcw_L1XE50, with a E_T^{miss} threshold at trigger level of 80 GeV. The tc_lcw stands for using topologically connected calorimeter clusters (topo-clusters), locally calibrated to the hadronic scale, as input to the trigger E_T^{miss} calculation. L1XE50 denotes the applied Level-1 trigger, which requires trigger-level E_T^{miss} above 50 GeV. Due to the increase in luminosity, the threshold for the lowest unprescaled E_T^{miss} trigger has been increased in 2016 and the trigger algorithm has changed: xe100_mht_L1XE50 was used in 2016. Here, mht indicates that the E_T^{miss} is calculated by summing over all anti-k_t jets with a radius of 0.4, calibrated at the electromagnetic scale.

The efficiencies of several E_T^{miss}-based triggers are compared in Fig. 9.3. Also the performance of E_T^{miss} triggers where muon information is added in the algorithm is shown but is not found to improve the efficiency significantly in the relevant regime above 200 GeV. Figure 9.4 shows the efficiency curves for HLT_xe80_tc_lcw_L1XE50, comparing 2015 and 2016 data. The behaviour of the trigger in the different datasets is comparable. The triggers are not fully efficient in all selection regions of the analysis: they are found to be approximately 95% efficient for events above the

9.2 Dataset

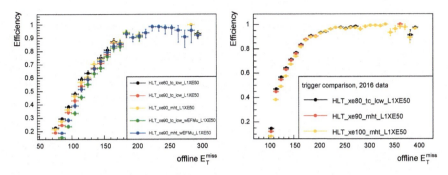

Fig. 9.3 Trigger efficiency of different E_T^{miss} triggers in data taken 2016 (left: 0.3 fb^{-1}, right: 3.5 fb^{-1})

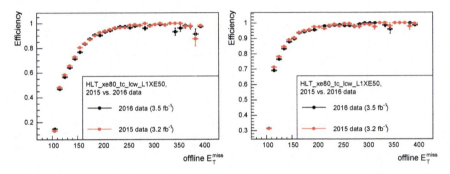

Fig. 9.4 Comparison of efficiency of the HLT_xe80_tc_lcw_L1XE50 E_T^{miss} trigger between 2015 data and 2016 data in the electron (left) and muon (right) channel

lowest E_T^{miss} cut applied in control regions of the analysis, $E_T^{miss} > 200$ GeV. Hence, the modelling of the efficiency curve below the plateau of full efficiency becomes important.

In Fig. 9.5, the data-MC agreement of the HLT_xe80_tc_lcw_L1XE50 efficiency is studied for 2015 data in three different scenarios: (a) with an inclusive preselection, (b) applying a b-jet veto to enrich W+jets and comparing to W+jets MC, and (c) requiring at least one b-jet to enrich the sample in $t\bar{t}$. Good agreement is observed in the regime above $E_T^{miss} > 200$ GeV for both electron and muon channel. The same comparison is presented in Fig. 9.6 for the HLT_xe100_mht_L1XE50 trigger efficiency and 2016 data. Again, good agreement is found in the regime above $E_T^{miss} > 200$ GeV.

The possibility of adding single-lepton triggers to the selection logic was investigated but the gain in signal acceptance was found to be very small (below 1%). Since a combination of E_T^{miss} and lepton triggers would add complexity but no significant improvement to the analysis, single-lepton triggers are only used for cross checks.

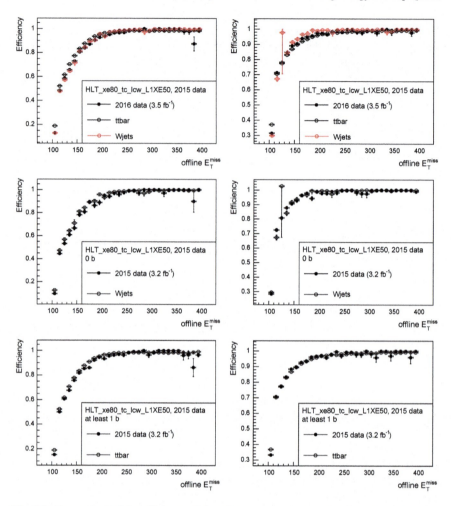

Fig. 9.5 Comparison of the efficiency of the `HLT_xe80_tc_lcw_L1XE50` trigger between 2015 data and MC with an inclusive preselection (top row), requiring zero b-jets (middle row) or at least one b-jet (bottom row). Electron channel (left) and muon channel (right) behave similarly

A data sample passing a single-photon trigger requiring a transverse momentum of $p_T > 140$ GeV is used to estimate the $t\bar{t}Z$ background contribution in a $t\bar{t}\gamma$ control region. This trigger is found to be more than 99% efficient for the relevant control region selection.

9.3 Event Reconstruction and Selection

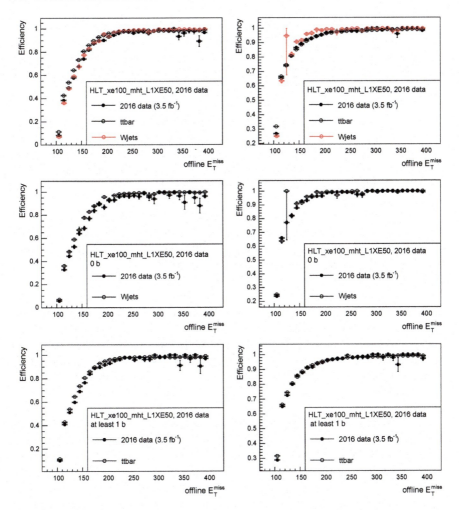

Fig. 9.6 Comparison of the efficiency of the HLT_xe100_mht_L1XE50 trigger between 2016 data and MC with an inclusive preselection (top row), requiring zero b-jets (middle row) or at least one b-jet (bottom row). Electron channel (left) and muon channel (right) behave very similarly

9.3 Event Reconstruction and Selection

9.3.1 Object Definition

Events are required to have at least one reconstructed vertex with two or more associated tracks above $p_T > 0.4$ GeV. If more than one vertex is found, the one with the highest sum of associated track p_T's is considered.

Leptons are defined in two categories. Leptons fulfilling basic quality and identification requirements ("baseline" leptons) enter the $E_\mathrm{T}^\mathrm{miss}$ calculation as well as the overlap removal procedure (described below) and are object to the veto of more than one lepton that features in the analysis. Tighter criteria are applied to define "signal" leptons, the leptons that are selected in the final state.

Baseline electrons are required to have $p_\mathrm{T} > 7$ GeV, $|\eta| < 2.47$, and to fulfil 'VeryLoose' likelihood identification criteria, described in Ref. [22]. Signal electrons have to pass the baseline selection and satisfy $p_\mathrm{T} > 25$ GeV, as well as the 'Loose' likelihood identification criteria. Their impact parameters with respect to the reconstructed primary vertex are constrained to $|z_0 \sin\theta| < 0.5$ mm and $|d_0|/\sigma_{d_0} < 5$, where σ_{d_0} denotes the uncertainty of d_0. Furthermore, signal electrons need to be isolated [27].

Baseline muons are selected via $p_\mathrm{T} > 6$ GeV, $|\eta| < 2.6$. They have to match the 'Loose' identification criteria described in Ref. [24]. Apart from passing the baseline requirements, signal muons are required to have $p_\mathrm{T} > 25$ GeV. Requirements on their impact parameters are given by $|z_0 \sin\theta| < 0.5$ mm and $|d_0|/\sigma_{d_0} < 3$. As signal electrons, signal muons need to be isolated.

An identified photon is required in the selection of the $t\bar{t} + \gamma$ sample that is used in the data-driven estimation of the $t\bar{t} + Z$ background. Photon candidates need to satisfy the 'Tight' identification criteria described in Ref. [24]. In addition, they are required to have $p_\mathrm{T} > 145$ GeV and $|\eta| < 2.37$. In order to ensure that the photon trigger is fully efficient for the selected events, the transition region between detector barrel and end-cap located between $1.37 < |\eta| < 1.52$ is excluded. Furthermore, photons must satisfy 'Tight' isolation criteria based on both track and calorimeter information.

As for the leptons, "baseline" jets fulfilling looser quality requirements are defined to enter the overlap removal and the $E_\mathrm{T}^\mathrm{miss}$ calculation, where "signal" jets are considered in the selection. Baseline jets are required to have $p_\mathrm{T} > 20$ GeV, signal jets must have $p_\mathrm{T} > 25$ GeV and $|\eta| < 2.5$. Signal jets with $p_\mathrm{T} < 60$ GeV have to pass further cuts that aim at rejecting jets originating from pileup [37]. Events containing a jet that does not pass specific jet quality requirements are vetoed from the analysis in order to suppress detector noise and non-collision backgrounds [27, 38].

Jets resulting from b-quarks (b-jets) are tagged using the MV2c10 b-tagging algorithm, which is based on quantities like impact parameters of associated tracks and reconstructed secondary vertices [28, 29]. A working point of 77% b-tagging efficiency is chosen.

Hadronically decaying tau leptons must fulfil the 'Loose' identification criteria described in Refs. [30, 31]. These tau candidates are required to have one or three associated tracks, with a total electric charge opposite to that of the selected electron or muon in the event. Furthermore, they are required to have $p_\mathrm{T} > 20$ GeV, and $|\eta| < 2.5$.

Apart from the identified (baseline) objects in the events, the *soft-term* enters the $E_\mathrm{T}^\mathrm{miss}$ calculation. For this analysis, it is constructed from track information: tracks that are associated with the primary vertex but not with the baseline physics objects are taken into account [32, 33]. For the photon selection, the calibrated photon

9.3 Event Reconstruction and Selection

is directly included in the E_T^{miss} calculation. Otherwise, photons and hadronically decaying tau leptons enter as jets, electrons, or via the soft-term.

Large-radius jets are clustered from all signal jets (small-radius $R = 0.4$) using the anti-k_t algorithm with $R = 1.0$ or 1.2. To reduce the impact of soft radiation and pileup, the large-radius jets are groomed using reclustered jet trimming, with a p_T fraction of 5% [34]. Electrons and muons are not included in the reclustering: the background acceptance would increase more than the signal efficiency. While large-radius jets are not directly considered in the overlap removal procedure, the signal jets from which they are reclustered, need to pass. The large-radius jet mass is used in the analysis. Its square is defined as the square of the four-vector sum of the momenta of the contained small-radius jets.

9.3.1.1 Overlap Removal

Detector signals might be interpreted and reconstructed as different physical objects. To avoid this double-labelling–and double-counting–of such objects, a so-called *overlap removal* procedure (OR), which was optimised for this analysis, is applied. A potential overlap is considered depending on shared tracks, ghost-matching [35] or radial distance of the objects, ΔR.

Electron/Muon: Some "Loose muon" objects are reconstructed including calorimetric information and can also be reconstructed as electrons. If an electron and a muon share a track, the electron is removed, except if the muon is based on calorimetric information.

Lepton/Jet: If an electron and a jet are closer than $\Delta R < 0.2$, the jet is removed, except if it is b-tagged. A jet is considered to overlap with a muon, if the muon can be ghost-matched to a nearby jet. The object is only reconstructed as a muon if the jet is not b-tagged and if it has less than three tracks above $p_T = 500$ MeV or if $p_T^{muon}/p_T^{jet} > 0.7$.

Jet/Lepton: If a jet, after the above steps, overlaps with a lepton in a cone of radius $R = 0.04 + 10/p_T^\ell$, up to a maximum radius of 0.4, the lepton is removed.

Taus and Photons: Taus are only used in the computation of the m_{T2}^τ variable to define a veto on them. If the event passes the tau veto, the tau object is no longer used and instead the jet object is considered for the rest of the computations. Photons are only used in the $t\bar{t} + \gamma$ control region. Jets overlapping with a photon in a cone with radius $R = 0.2$ are removed. In the rest of the regions photons are not considered and overlapping photon/jets are always treated as jets. If an electron, after the previous steps, overlaps with a tau candidate or a photon in a cone with radius $R = 0.1$, the electron is taken.

9.4 Discriminating Variables

All processes containing a leptonically decaying W boson can be reduced effectively by reconstructing the transverse mass of the common parent particle of the lepton and the E_T^{miss}, if a cut is put above the W boson mass. The transverse mass m_T is defined as follows:

$$m_T = \sqrt{2 \cdot p_T^\ell \cdot E_T^{\text{miss}} \left(1 - \cos \Delta\Phi(\vec{\ell}, \vec{E}_T^{\text{miss}})\right)}. \tag{9.1}$$

Here, p_T^ℓ is the lepton p_T, and $\Delta\phi(\vec{\ell}, \vec{E}_T^{\text{miss}})$ is the azimuthal angle between the lepton and the $\vec{E}_T^{\text{miss}}$ direction. It is assumed that the lepton mass is negligible.

By this mean, single-leptonic $t\bar{t}$ and W+jets processes are suppressed by about 90%. Events originating from $t\bar{t}$ and W+jets can escape such a cut either due to the limited resolution of the reconstructed m_T or if an additional source of E_T^{miss} is present in the event. The latter is true for signal events, but also for dileptonic $t\bar{t}$ events where one lepton is not identified or out of acceptance. Also $t\bar{t}$ events in which one W decays leptonically and one into a hadronic tau often have a larger m_T. Hence, most of the analysis-specific variables, described in further detail in Ref. [36], aim at rejecting these background components.

A first strategy is to try to reconstruct the hadronic top candidate. Via a χ^2-minimisation, the three jets that are best compatible with originating from a top quark are selected, according to their momenta and considering their momentum resolution. Their invariant mass is defined to be m_{top}^χ. In the case of dileptonic $t\bar{t}$ this variable will not be close to the top mass by any means, whereas this is the case for signal and background events containing a true hadronic top decay. This approach is extended to the case of dileptonic $t\bar{t}$ with a lost lepton by the so-called *topness* variable [37]. Furthermore, the E_T^{miss} perpendicular to the reconstructed leptonic top candidate can be used to distinguish between signal and background. After the hadronic top candidate is reconstructed as described above, the additional b-jet is combined with the identified lepton to form the leptonic top. While this $E_{T,\perp}^{\text{miss}}$ is expected to be small for background events, where the E_T^{miss} from the neutrino is aligned with the leptonic top, this variable is likely large for signal events (see Fig. 9.7).

A second strategy, based on the m_T approach, is to reconstruct the different decay chains and give an upper bound on the hypothetical parent mass. First, the so-called *stransverse mass*, m_{T2} [38] can be defined. It extends the transverse mass m_T to decay topologies with two branches, a and b, originating from the same, pair-produced parent A. In each branch, it is assumed that there are some particles with measured momenta and some unmeasured particles. The sum of the measured momenta in branch $i \in \{a, b\}$ is denoted $p_i = (E_i, \vec{p}_{Ti}, p_{zi})$ and the sum of the unmeasured momenta is denoted $q_i = (F_i, \vec{q}_{Ti}, q_{zi})$. Then, $m_{p_i}^2 = E_i^2 - \vec{p}_i^2$ and $m_{q_i}^2 = F_i^2 - \vec{q}_i^2$. The m_T of the particles in branch i is given by:

$$m_{Ti}^2 = \left(\sqrt{p_{Ti}^2 + m_{p_i}^2} + \sqrt{q_{Ti}^2 + m_{q_i}^2}\right)^2 - (\vec{p}_{Ti} + \vec{q}_{Ti})^2 \tag{9.2}$$

9.4 Discriminating Variables

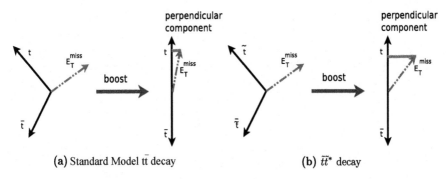

Fig. 9.7 Sketch of the definition of the perpendicular E_T^{miss} variable

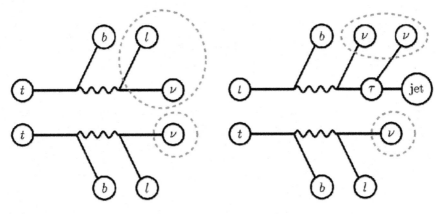

Fig. 9.8 Schematic view on the variables am_{T2} (left) and m_{T2}^{τ} (right)

which, in the case of $m_{q_i} = m_{p_i} = 0$, is the same as the expression for m_T given above. It has an end point at the parent mass m_A. Now, m_{T2}, is defined as a minimisation over the allocation of $\vec{p}_T^{\text{miss}}$ between $\vec{q}_{Ta}$ and $\vec{q}_{Tb}$ of the maximum of the corresponding m_{Ta} or m_{Tb}:

$$m_{T2} \equiv \min_{\vec{q}_{Ta}+\vec{q}_{Tb}=\vec{p}_T^{\text{miss}}} \{\max(m_{Ta}, m_{Tb})\}. \tag{9.3}$$

An assumption of m_{q_a} and m_{q_b} must be made in the computation of m_{Ta} and m_{Tb}. The result of the above minimisation is the minimum parent mass m_A consistent with the observed kinematic distributions under the inputs m_{q_a} and m_{q_b}. The variants of m_{T2} described below only differ in the considered measured particles, assumed unmeasured particles, and choices for the input masses, m_{q_a} and m_{q_b}.

The m_{T2} is adapted to target dileptonic $t\bar{t}$ by accounting for missed leptons in the so-called asymmetric m_{T2} (am_{T2}) [39–42] (see Fig. 9.8). Here, decay branch a assumes a b-jet as visible object and takes the lepton originating from the leptonically-decaying W boson as lost, and so the lepton and the neutrino as unmeasured. Hence,

m_{q_a} equals the W mass. Decay branch b takes a b-jet[6] and the lepton as visible objects, the neutrino from the leptonically-decaying W boson is the invisible part. Both possible assignments of the b-jets are tested and the one resulting in the smaller am_{T2} is taken. In such a dileptonic $t\bar{t}$ scenario, the am_{T2} is bounded from above by the top quark mass, while signals typically exceed this bound.

Similarly, the m_{T2}^τ variable is optimised for the case of a W boson that decays into a hadronic tau (see Fig. 9.8). Decay branch a considers a reconstructed hadronic tau candidate as visible object, the two neutrinos resulting from the W and the tau decay present the unmeasured components. Decay branch b takes the lepton as visible and its neutrino as invisible object. Both m_{q_a} and m_{q_b} are taken to be zero. This variable can be used as a tau veto for $t\bar{t}$ decays into one lepton (electron or muon) and one hadronically-decaying tau lepton by removing events in which the m_{T2}^τ does not exceed 80 GeV, i.e. the W mass bound.

It proved useful to define $H_{T,\text{sig}}^{\text{miss}}$, the significance of a purely object-based missing transverse jet energy, as:

$$H_{T,\text{sig}}^{\text{miss}} = \frac{|\vec{H}_T^{\text{miss}}| - M}{\sigma_{|\vec{H}_T^{\text{miss}}|}}, \qquad (9.4)$$

where $\vec{H}_T^{\text{miss}}$ is the negative sum of the jet and lepton momentum vectors. The denominator gives the resolution of $\vec{H}_T^{\text{miss}}$ and considers the per-event energy resolution of the jets determined from the per-event jet energy uncertainties. The lepton energy is assumed to be measured significantly better and hence its resolution is neglected. The parameter M denotes a "characteristic scale" of the background [43]. Based on optimisation studies it is fixed at 100 GeV in this analysis.

9.5 Signal Regions

9.5.1 Preselection

A preselection, common to all signal and control regions is applied to select the events that are considered for the different aspects of the analysis. The one-lepton final state is selected by requiring exactly one identified signal lepton and vetoing additional baseline leptons. Lepton here includes only electrons and muons. At least two signal jets present in the event are required as well. Since all signal scenarios involve invisible particles, a E_T^{miss} of at least 200 GeV is required. In case this is not fulfilled, but a photon is found with a p_T above 200 GeV the event can still be considered for the $t\bar{t}\gamma$ control region. In order to further reduce the contribution from multijet events that feature a wrongly-identified lepton and fake E_T^{miss} due to jet energy mis-measurements, the transverse mass between the signal lepton and the

---

[6]In case there is only one or more than two b-jets found in the event, the ones with the highest b-tagging weights are considered.

9.5 Signal Regions

Table 9.2 Common preselection for the optimisation of the signal regions

Selection	Comments
`HLT_xe80_tc_lcw_L1XE50` trigger jet cleaning	Veto events that contain a jet that fails the loose jet cleaning criteria
Exactly one signal lepton	And no additional baseline leptons
≥ 4 signal jets	Reduce low-jet multiplicity backgrounds (diboson, W/Z)
$E_T^{miss} > 200$ GeV	Start of the XE trigger plateau
$m_T > 30$ GeV	Control of QCD/multijets
$\lvert \Delta\phi(j_{1,2}, \vec{p}_T^{\,miss}) \rvert > 0.4$	Control of QCD multijet backgrounds
m_{T2}^τ based τ-veto ($m_{T2}^\tau > 80$ GeV)	Remove events with hadronic tau candidates

$\vec{p}_T^{\,miss}$ has to be larger than 30 GeV. Furthermore, the angular separation between $\vec{p}_T^{\,miss}$ and each of the two leading jets has to exceed 0.4. Also the requirement of $H_{T,sig}^{miss} > 5$ reduces the amount of multijet events entering the selection. After these selection cuts the multijet background is found to be negligible. The preselection is summarised in Table 9.2 and selected distributions are shown in Fig. 9.9. Data and Monte Carlo are in reasonable agreement in the bulk of the distributions, while some deviations appears e.g. in the tail of am_{T2}. Further, the data overshoots the Monte Carlo for very low values of E_T^{miss}, m_T and $H_{T,sig}^{miss}$, where a contribution from multijet backgrounds is expected. As discussed, after applying the above requirements on m_T and $H_{T,sig}^{miss}$ this contribution becomes negligible.

9.5.2 Dark Matter Optimisation

While the previous results on top-associated DM production were obtained from a reinterpretation of one of the stop signal regions, the selection was now optimised for sensitivity to Dark Matter signals, taken from the DM Simplified Model introduced above. A variety of different variables that can possibly discriminate between signal and background were tested to find an optimal selection. Since the integrated luminosity available for the final analysis was not known, 7 fb^{-1} have been assumed. Small changes in the selections would be expected if the optimisation was performed using the actual luminosity of 13 fb^{-1}.

9.5.2.1 Optimisation Procedure

The above introduced preselection is assumed before the optimisation is performed (see Table 9.2). The backgrounds were taken from MC without applying any normalisation factor. A possible improvement of this procedure would be to apply approximate normalisation factors obtained in previous rounds of the analysis already here. For example, the $t\bar{t}Z$ background is found to have a normalisation factor much above

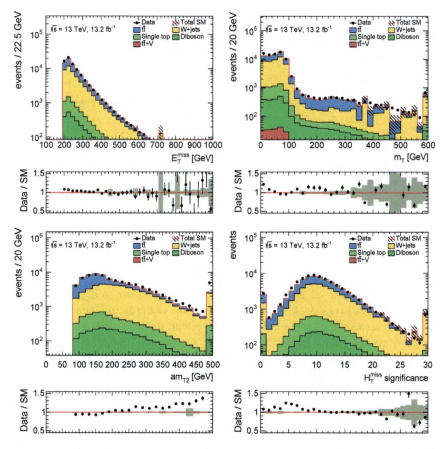

Fig. 9.9 Comparison of data and simulation at pre-selection before the fit to data for E_T^{miss} (top left), m_T (top right), am_{T2} (bottom left) and $H_{T,sig}^{miss}$ (bottom right). Only statistical uncertainties are displayed. The last bin includes the overflow

one, as will be discussed below, which alters the performance of the optimised SRs. A flat uncertainty of 20% was assumed for all background contributions to approximate the final uncertainty expected to be obtained. The standard object definitions listed in Sect. 9.3 are used.

The following significance estimator was considered as a figure of merit:

$$\sigma = \frac{n_{sig}}{\sqrt{n_{bkg} + 1 + \sigma_{bkg}^2}}, \qquad (9.5)$$

where σ_{bkg} denotes the absolute uncertainty on the background estimate. The results were alternatively cross-checked using a binomial z-score [44] as an estimator as done for the other SRs. The resulting selections were in good agreement.

9.5 Signal Regions

The optimisation is performed iteratively. In each step, the variables are ranked and an optimal cut value is found according to the significance estimator. The best performing cut is then applied and the procedure is repeated until no significant improvement is achieved by adding another cut. Since many of the tested variables are expected to be correlated, a damping function is applied: during the first iterations of the procedure, the background efficiency needs to stay above a certain threshold that depends on the iteration. In the first step, at least 50% of the backgrounds are required to pass the cut. This threshold is multiplied by one half at each iteration. The same variable can be ranked highest multiple times. By applying this damping, the procedure is much less sensitive to statistical fluctuations and the correlation effects between variables is moderated.

9.5.2.2 Signal Benchmarks

For the optimisation, two different benchmark points of the DM Simplified Model have been chosen:

- assume a natural coupling of $g_{q,\chi} = 1$ and a relatively light mediator of $m_\phi = 100$ GeV,
- assume a maximal coupling of $g_{q,\chi} = 3.5$ and a heavier mediator of $m_\phi = 350$ GeV.

The DM particle, is taken to be light ($m_\chi = 1$ GeV). Going to higher DM masses would only have a small effect as long as the mediator mass allows to produce the DM particles on-shell. Furthermore, it would almost exclusively affect the cross section and not the signal acceptance, since the observed kinematic behaviour would not change in the on-shell regime. The benchmarks, $(m_\phi, m_\chi, g_{q,\chi}) = (100$ GeV, 1 GeV, 1$)$ and $(350$ GeV, 1 GeV, 3.5$)$, are studied both for scalar and pseudo-scalar mediators.

9.5.2.3 Results

The optimisation revealed that a SR targeting the $(m_\phi, m_\chi, g_{q,\chi}) = (350$ GeV, 1 GeV, 3.5$)$ benchmark performs almost as well for a wide range of mediator masses (300–450 GeV) than a dedicated optimisation. Furthermore, the pseudo-scalar signature is found to be well covered by the selections optimised on scalar signals both for a mediator mass of 100 and 350 GeV, accepting a loss in significance of up to 20%. Hence, two SRs are defined for the DM signals: one region targeting lower mediator masses and smaller couplings ("DM_low"), and one targeting higher mediator masses and larger couplings ("DM_high"). In Table 9.3, the resulting cuts for the two DM SRs are given.

In Fig. 9.10, several distributions are shown for the DM_low selection. All DM_low cuts but the one on the displayed quantity are applied. The vertical line

Table 9.3 Overview of the event selections defining the two DM signal regions. The common event preselection as defined in Table 9.2 is applied in all cases

Variable	DM_low	DM_high
Number of jets	≥ 4	≥ 4
Leading jet p_T	>60 GeV	>50 GeV
Second jet p_T	>60 GeV	>50 GeV
Third jet p_T	>40 GeV	>50 GeV
E_T^{miss}	>300 GeV	>330 GeV
$H_{T,sig}^{miss}$	>14	>9.5
m_T	>120 GeV	>220 GeV
am_{T2}	>140 GeV	>170 GeV
$\Delta\phi(\vec{p}_T^{\,miss}, \ell)$	>0.8	–
$\min(\Delta\phi(\vec{p}_T^{\,miss}, \text{jet}_i))$	>1.4	>0.8
Number of b-tags	≥ 1	≥ 1

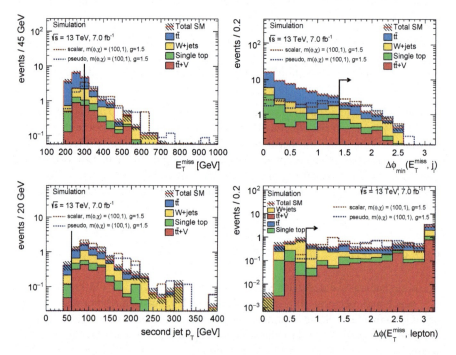

Fig. 9.10 The E_T^{miss} (top left), $\min(\Delta\phi(\vec{p}_T^{\,miss}, \text{jet}_i))$ (top right), second leading jet p_T (bottom left) and $\Delta\phi$ between the E_T^{miss} and the lepton (bottom right) distributions after applying all DM_low requirements but the one on the shown distribution. The cut which would be applied on this distribution is shown by the vertical line. The overview of the selection is given in Table 9.3

9.5 Signal Regions

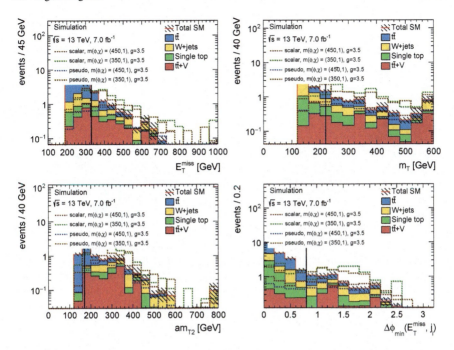

Fig. 9.11 The E_T^{miss} (top left), m_T (top right), am_{T2} (bottom left) and $\min(\Delta\phi(\vec{p}_T^{miss}, \text{jet}_i))$ (bottom right) distributions after applying all DM_high requirements but the one on the shown distribution. The cut which would be applied on this distribution is shown by the vertical line. The overview of the selection is given in Table 9.3

in the plot shows where the cut on this distribution would be applied. The distributions are shown for all relevant MC processes and different DM signal samples. Analogously, Fig. 9.11 shows the distributions for the DM_high selection.

9.5.2.4 Expected Performance

Figure 9.12 shows the background composition and expected fraction of signal events in the two DM regions. In DM_low, dileptonic $t\bar{t}$ makes up over 30% of the background, as well as W+jets, while $t\bar{t}Z$ presents the third-largest background contribution. Almost 40% of the events in this region would be expected to come from a signal from a scalar mediator and $(m_\phi, m_\chi, g_{q,\chi}) = (100 \text{ GeV}, 1 \text{ GeV}, 1)$. DM_high would see around 65% of its events coming from a signal from a scalar mediator and $(m_\phi, m_\chi, g_{q,\chi}) = (350 \text{ GeV}, 1 \text{ GeV}, 3.5)$. Here, dileptonic $t\bar{t}$ and $t\bar{t}Z$ contribute most to the expected background, both at the level of 30%.

Figure 9.13 shows the expected discovery significance as a function of the integrated luminosity for the two benchmark points for a scalar mediator, $(m_\phi, m_\chi, g_{q,\chi}) = (100 \text{ GeV}, 1 \text{ GeV}, 1)$ and $(m_\phi, m_\chi, g_{q,\chi}) = (350 \text{ GeV}, 1 \text{ GeV}, 3.5)$. For the signal

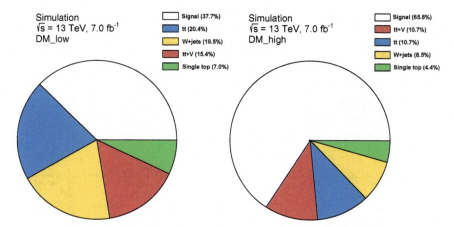

Fig. 9.12 Composition of the DM_low (left) and DM_high (right) signal regions. The benchmark scalar mediator signals in DM_low (left): $(m_\phi, m_\chi, g_{q,\chi}) = (100\ \text{GeV}, 1\ \text{GeV}, 1)$ and DM_high (right): $(m_\phi, m_\chi, g_{q,\chi}) = (350\ \text{GeV}, 1\ \text{GeV}, 3.5)$ are included as well as the different background contributions

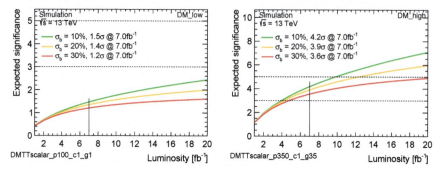

Fig. 9.13 Expected discovery significance as function of luminosity for the benchmark scalar mediator signals in DM_low (left): $(m_\phi, m_\chi, g_{q,\chi}) = (100\ \text{GeV}, 1\ \text{GeV}, 1)$ and DM_high (right): $(m_\phi, m_\chi, g_{q,\chi}) = (350\ \text{GeV}, 1\ \text{GeV}, 3.5)$. Flat uncertainties on the background event yield of $\sigma_b = 10, 20, 30\%$ are considered

$(m_\phi, m_\chi, g_{q,\chi}) = (100\ \text{GeV}, 1\ \text{GeV}, 1)$ in DM_low the significance does not reach 3σ even for 20 fb^{-1}. However, signal points with slightly increased coupling strengths could be observed at the 3σ level with approximately 10 fb^{-1}. For the signal $(m_\phi, m_\chi, g_{q,\chi}) = (350\ \text{GeV}, 1\ \text{GeV}, 3.5)$ in DM_high a significance over 3σ is expected to be reached for a dataset between 4 and 5 fb^{-1}.

9.5 Signal Regions

Table 9.4 Overview of the event selections for the seven SRs considered in the analysis. Round brackets are used to describe lists of values and square brackets denote intervals

Common event selection	
Trigger	E_T^{miss} trigger
Lepton	Exactly one signal lepton (e, μ), no additional baseline leptons
Jets	At least two signal jets, and $|\Delta\phi(\mathrm{jet}_i, \vec{p}_T^{\,miss})| > 0.4$ for $i \in \{1, 2\}$
Hadronic τ veto	Veto events with a hadronic tau candidate and $m_{T2}^\tau < 80$ GeV

Variable	SR1	tN_high
Number of (jets, b-tags)	($\geq 4, \geq 1$)	($\geq 4, \geq 1$)
Jet $p_T >$ (GeV)	(80 50 40 40)	(120 80 50 25)
E_T^{miss} (GeV)	>260	>450
$E_{T,\perp}^{miss}$ (GeV)	–	>180
$H_{T,sig}^{miss}$	>14	>22
m_T (GeV)	>170	>210
am_{T2} (GeV)	>175	>175
topness	>6.5	–
m_{top}^χ (GeV)	<270	–
$\Delta R(b, \ell)$	<3.0	<2.4
Leading large-R jet p_T (GeV)	–	>290
Leading large-R jet mass (GeV)	–	>70
$\Delta\phi(\vec{p}_T^{\,miss},$ 2nd large-R jet)	–	>0.6

Variable	bC_diag	bC_med	bCbv
Number of (jets, b-tags)	($\geq 4, \geq 2$)	($\geq 4, \geq 2$)	($\geq 2, =0$)
Jet $p_T >$ (GeV)	(70 60 55 25)	(170 110 25 25)	(120 80)
b-tagged jet $p_T >$ (GeV)	(25 25)	(105 100)	–
E_T^{miss} (GeV)	>230	>210	>360
$H_{T,sig}^{miss}$	>14	>7	>16
m_T (GeV)	>170	>140	>200
am_{T2} (GeV)	>170	>210	–
$|\Delta\phi(\mathrm{jet}_i, \vec{p}_T^{\,miss})|(i=1)$	>1.2	>1.0	>2.0
$|\Delta\phi(\mathrm{jet}_i, \vec{p}_T^{\,miss})|(i=2)$	>0.8	>0.8	>0.8
Leading large-R jet mass (GeV)	–	–	[70, 100]
$\Delta\phi(\vec{p}_T^{\,miss}, \ell)$	–	–	>1.2

Variable	DM_low	DM_high
Number of (jets, b-tags)	($\geq 4, \geq 1$)	($\geq 4, \geq 1$)
Jet $p_T >$ (GeV)	(60 60 40 25)	(50 50 50 25)
E_T^{miss} (GeV)	>300	>330
$H_{T,sig}^{miss}$	>14	>9.5
m_T (GeV)	>120	>220
am_{T2} (GeV)	>140	>170
$\min(\Delta\phi(\vec{p}_T^{\,miss}, \mathrm{jet}_i))$ ($i \in (1-4)$)	>1.4	>0.8
$\Delta\phi(\vec{p}_T^{\,miss}, \ell)$	>0.8	–

9.5.3 Signal Region Overview

After the preselection detailed above, $t\bar{t}$ and W+jets processes represent the most dominant backgrounds. Both of these backgrounds can be reduced by a cut on the transverse mass of lepton and $\vec{p}_T^{\text{miss}}$. Furthermore, if a hadronic tau candidate is found in the event, the variable m_{T2}^{τ} is required to be larger than the W boson mass to reject $t\bar{t}$ events where one W boson decays into a hadronic tau. Apart from the DM SRs that have already been introduced, two SRs targeting tN signal scenarios are defined. SR1 targets moderate stop masses, whereas tN_high is optimised for very high stop masses, where the decay products are expected to be highly boosted. It therefore relies on large-radius jets. SR1 is inherited from the previous publication of this analysis [8], which saw a mild excess in this region that should be reviewed with more data.

The bC signal scenario is covered by three SRs: bC2x_diag and bC2x_med target the scenario of $m_{\tilde{\chi}^\pm} = 2m_{\tilde{\chi}^0}$ with small and medium mass differences between stop and chargino, respectively. Their selection relies on high-p_T b-jets. A small mass splitting between the stop and the chargino of 10 GeV is assumed in the optimisation of the so-called bCbv SR. Here, the b-jets are expected to be too soft to be identified and hence a b-jet veto is applied.

The SR definitions are summarised in Table 9.4. Note that the selections are not orthogonal and the overlap between the different SRs can be significant.

9.6 Background Estimation

The dominant contributions to the background entering the signal selections stem from $t\bar{t}$, single top Wt, $t\bar{t} + Z(\rightarrow \nu\bar{\nu})$, and W+jets processes. Since the semi-leptonic component of $t\bar{t}$ can be efficiently reduced, mostly dileptonic $t\bar{t}$ events where one lepton is out of acceptance or not identified and hence escaping the veto, as well as $t\bar{t}$ events featuring one lepton and one hadronic tau in the final state, remain in the SRs. Small backgrounds come from diboson, $t\bar{t}W$ and Z+jets events. As discussed above, the multijet background is found to be negligible after preselection.

In order to estimate the major backgrounds, B = {$t\bar{t}$, W+jets, single-top, $t\bar{t}Z$}, dedicated control regions enriched in the respective background processes, are defined: CR = {TCR, WCR, STCR, TZR}, respectively. They are designed to be kinematically as similar as possible to the SRs. The inversion of few specific cuts makes the control regions orthogonal to the SRs, reduces the possible signal contribution and enhances the yield and purity of the background in question. Minor backgrounds, b = {Z+jets, dibosons}, are purely taken from MC simulation.

The *normalisation factor* n_i for any major background i is defined as follows:

$$n_i = N_{CR_i}^{i,data} / N_{CR_i}^{i,MC}, \tag{9.6}$$

9.6 Background Estimation

with:

$$N_{CR_i}^{i,data} = N_{CR_i}^{data} - \sum_{j \neq i, j \in B} n_j \cdot N_{CR_i}^{j,MC} - \sum_{k \in b} N_{CR_i}^{k,MC}. \quad (9.7)$$

Here, $N_{CR_i}^{i,MC}$ denotes the Monte Carlo event yield for the background i in control region in question, and $N_{CR_i}^{data}$ denotes the total number of data events observed in this control region. All normalisation factors are determined in a simultaneous likelihood fit [45] to all control regions of one SR, for each SR. The findings are then used in the prediction of background events in the SR via *transfer factors* t_i for each background i:

$$t_i = N_{SR}^{i,MC} / N_{CR_i}^{i,MC}. \quad (9.8)$$

The total number of expected events in a SR is then given by:

$$N_{SR} = \mu_{sig} \cdot N^{sig} + \sum_{i \in B, b} N_{SR}^i = \mu_{sig} \cdot N^{sig} + \sum_{j \in B} n_j \cdot t_j \cdot N_{CR_j}^{j,MC} + \sum_{k \in b} N_{SR}^{k,MC}. \quad (9.9)$$

The signal strength μ_{sig} is used to vary or constrain the assumed signal contribution. For the determination of the normalisation factors, the above formula is applied to the control regions and the signal strength is set to zero (*background-only fit*). The four fit parameters, namely the normalisations of $t\bar{t}$, single top, W+jets, and $t\bar{t} + W/Z$ are affected by systematic uncertainties, which are treated as Gaussian nuisance parameters in the fit.

The resulting background modelling is verified in validation regions. An overview of the control and validation regions considered is given in Fig. 9.14. The definitions of the control and validation regions for the DM SRs is given in Table 9.5. For the other SRs such an overview can be found in Appendix C.

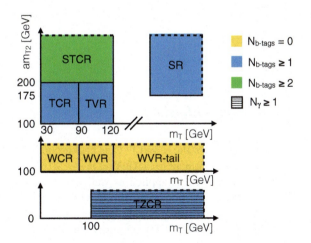

Fig. 9.14 A schematic diagram for the various event selections used to estimate and validate the background normalizations. Solid lines indicate kinematic boundaries while dashed lines indicate that the events can extend beyond the boundary. CR, VR, and SR stand for control, validation, and signal region, respectively. T, ST, TZ, and W stand for $t\bar{t}$, single top, $t\bar{t} + Z$, and W+jets, respectively

Table 9.5 Overview of the event selections for DM signal regions and the associated $t\bar{t}$ (TCR), W+jets (WCR), and Wt (STCR) control regions. Round brackets are used to describe lists of values and square brackets denote intervals

Common event selection for DM	
Trigger	E_T^{miss} trigger
Lepton	Exactly one signal lepton (e, μ), no additional baseline leptons
Jets	At least two signal jets, and $|\Delta\phi(\text{jet}_i, \vec{p}_T^{\text{miss}})| > 0.4$ for $i \in \{1, 2\}$
Hadronic τ veto	Veto events with a hadronic tau decay and $m_{T2}^\tau < 80$ GeV

Variable	DM_low	TCR/WCR	STCR
≥ 4 jets with $p_T >$ (GeV)	(60 60 40 25)	(60 60 40 25)	(60 60 40 25)
E_T^{miss} (GeV)	>300	>200/>230	>200
$H_{T,\text{sig}}^{\text{miss}}$	>14	>8	>8
m_T (GeV)	>120	[30, 90]	[30, 120]
am_{T2} (GeV)	>140	[100, 200]/>100	>200
$\min(\Delta\phi(\vec{p}_T^{\text{miss}}, \text{jet}_i))$ ($i \in \{1-4\}$)	>1.4	>1.4	>1.4
$\Delta\phi(\vec{p}_T^{\text{miss}}, \ell)$	>0.8	>0.8	–
$\Delta R(b_1, b_2)$	–	–	>1.8
Number of b-tags	≥ 1	$\geq 1/=0$	≥ 2

Variable	DM_high	TCR/WCR	STCR
≥ 4 jets with $p_T >$ (GeV)	(50 50 50 25)	(50 50 50 25)	(50 50 50 25)
E_T^{miss} (GeV)	>330	>300/>330	>250
$H_{T,\text{sig}}^{\text{miss}}$	>9.5	>9.5	>5
m_T (GeV)	>220	[30, 90]	[30, 120]
am_{T2} (GeV)	>170	[100, 200]/>100	>200
$\min(\Delta\phi(\vec{p}_T^{\text{miss}}, \text{jet}_i))$ ($i \in \{1-4\}$)	>0.8	>0.8	>0.8
$\Delta R(b_1, b_2)$	–	–	>1.2
Number of b-tags	≥ 1	$\geq 1/=0$	≥ 2

9.6.1 Control Regions

Control regions for the $t\bar{t}$ and W+jets backgrounds are defined for all SRs by lowering the m_T requirement to 30 GeV $< m_T <$ 90 GeV. For the W+jets control region (WCR), a b-jet veto is applied as well. For the $t\bar{t}$ control region (TCR) an upper cut on am_{T2} avoids potential overlap with the single top control region (described below). Individual selection cuts are loosened to allow for sufficient statistics in the control regions. In the TCRs a $t\bar{t}$ purity between 51 and 91% is achieved, the W+jets purity in the WCRs is around 75%.

9.6 Background Estimation

For the single top background, the definition of a control region is not straightforward, since $t\bar{t}$ events may easily enter the selection due to the similar characteristics of $t\bar{t}$ and Wt. However, by applying the following cuts, the Wt purity can be enhanced. Requiring two b-jets reduces the W+jets contamination. Since the mass of a Wb system that does not origin from a top is typically higher than for an on-shell top quark in the selected phase space, a cut on $am_{T2} > 200\,\text{GeV}$ is effective in reducing the $t\bar{t}$ contamination. The remaining $t\bar{t}$ can evade this cut if a c-quark from the W decay gets b-tagged and represents one of the two b-jets entering the am_{T2} calculation. In this case, the radial distance between these "b-jets" is typically smaller than in real Wt events. Hence, a cut on $\Delta R(b_1, b_2) > 1.2$ is introduced. This is further tightened in the DM_low and bC_diag regions, where this is possible without reducing the statistics too much. With this strategy, control regions with a Wt purity of up to 50% are obtained, which provide a reasonable handle on this background that is especially relevant for the bC2x regions.

The $t\bar{t} + Z(\rightarrow \nu\bar{\nu})$ background constitutes a significant fraction of the total background, especially in the SRs relying on tight E_T^{miss} cuts, like tN_high and the DM regions. While the definition of a tri-lepton region in which the Z boson decays to charged leptons was used as a cross-check, the integrated luminosity of the dataset does not allow to use it as a control region. A $t\bar{t}\gamma$ control region is designed instead to constrain the $t\bar{t}Z$ background from data. There are two major differences between $t\bar{t}Z$ and $t\bar{t}\gamma$ that also lead to different kinematic characteristics. The finite mass of the Z boson as opposed to the massless photon becomes less relevant the more the boson p_T exceeds the Z boson mass. Since all SRs require high E_T^{miss}, the Z boson–and the corresponding photon–must have high p_T, which was verified to be high enough to moderate the impact of the Z boson mass. The second difference is that additional bremsstrahlung from the Z boson is suppressed at LHC energies, while in a significant fraction of $t\bar{t}\gamma$ events the photon is radiated off one of the top quarks. Also this difference is much smaller for high boson p_T. It is taken into account in the simulations and hence is also considered in the background estimate. For the control region selection, a high-p_T photon above 200 GeV is required in addition to the one-lepton selection. It is treated as invisible for all variable calculations: the highest-p_T photon is added vectorially to $\vec{p}_T^{\text{miss}}$ and this sum is used to construct $\tilde{E}_{T,\text{corr}}^{\text{miss}} = |\vec{p}_T^{\text{miss}} + \vec{p}_T^{\gamma}|$, $\tilde{m}_T$, and $\tilde{H}_{T,\text{sig}}^{\text{miss}}$. The jet p_T thresholds are kept identical to those in the SRs to minimise the impact of systematic uncertainties on the jet energy scale (JES). In order to make this control region kinematically more similar to the SRs, cuts on the modified E_T^{miss}, m_T and $H_{T,\text{sig}}^{\text{miss}}$ are applied. An upper cut of $E_T^{\text{miss}} < 200\,\text{GeV}$ forces this control region to be orthogonal to other regions. The selection leads to a $t\bar{t}\gamma$ sample with over 90% purity, where the largest contamination comes from $W\gamma$ + jets. The total number of data events exceeds the MC prediction by 30–47%, while no apparent mis-modelling of relevant variables is observed.

The contribution from the multijet background was estimated using a data-driven procedure in context of the previous release of this analysis [8]. It was found to be negligible. In this analysis, the amount of multijet background entering the selections is confirmed to be negligible in a multijet-enriched region, defined by lowering the

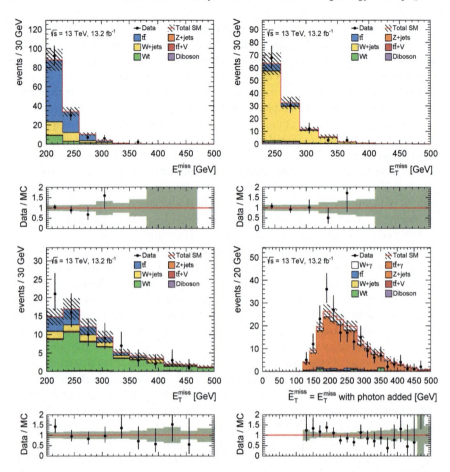

Fig. 9.15 Distributions of E_T^{miss} (top left for TCR, top right for WCR and bottom left for STCR), and photon-corrected E_T^{miss} (bottom right for TZCR) for events in the CRs associated with DM_low where each background ($t\bar{t}$, W+jets, Wt, and $t\bar{t} + W/Z$) is normalised by normalisation factors obtained in a background-only fit. The uncertainty band includes statistical and experimental systematic uncertainties. The last bin includes overflow

E_T^{miss} down to 50 GeV: when requiring E_T^{miss} above 200 GeV, basically no multijet background remains. All other small backgrounds are determined from simulation, with the cross sections normalised to the most accurate theoretical prediction available.

The E_T^{miss} distribution for the CRs associated with DM_low are shown in Fig. 9.15 and in Fig. 9.16 for DM_high. The distributions in data are found to be well modelled in all control regions.

9.6 Background Estimation

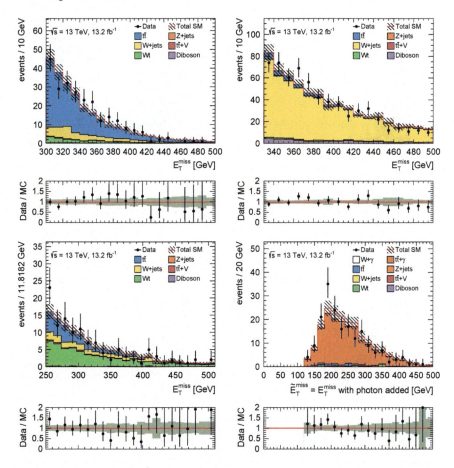

Fig. 9.16 Distributions of E_T^{miss} (top left for TCR, top right for WCR and bottom left for STCR), and photon-corrected E_T^{miss} (bottom right for TZCR) for events in the CRs associated with DM_high where each background ($t\bar{t}$, W+jets, Wt, and $t\bar{t} + W/Z$) is normalised by normalisation factors obtained in a background-only fit. The uncertainty band includes statistical and experimental systematic uncertainties. The last bin includes overflow

9.6.2 Validation Regions

The normalisation factors determined by the fit and the resulting background modelling is verified in dedicated validation regions (VRs). These regions are built to be orthogonal to both signal and control regions to provide a statistically independent test of the fit results. The possible signal contamination in these regions was found to be at most 10%. For the validation of the $t\bar{t}$ and W+jets background estimate, one VR for each process and SR gets defined in the same way as the respective control region but within an m_T window of $90\,\text{GeV} < m_T < 120\,\text{GeV}$. This strategy cannot be applied in the case of the single top region, since there the m_T cut needed to be

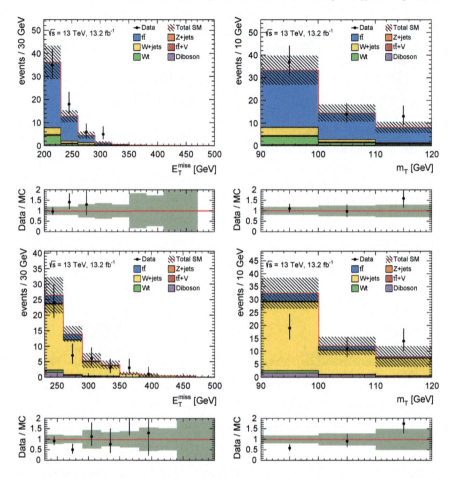

Fig. 9.17 Distributions of E_T^{miss} (left) and m_T (right) for the TVR (top) and WVR (bottom) selection associated with DM_low. Each background ($t\bar{t}$, W+jets, Wt, and $t\bar{t} + W/Z$) is normalised according to the result of the background-only fit to the control regions. Statistical and experimental systematic uncertainties are included in the error band

extended up to 120 GeV in order to retain a sufficient amount of events passing the selection. Therefore, no single top VR could be defined.

The agreement of data and MC prediction in the $t\bar{t}$ and W+jets validation regions of DM_low and DM_high is presented in Figs. 9.17 and 9.18, respectively. The distributions of E_T^{miss} and m_T are found to agree reasonably well, although a trend in the ratio is observed for m_T in the DM_low WVR, which slightly exceeds the uncertainty band in the low-m_T bin.

9.6 Background Estimation

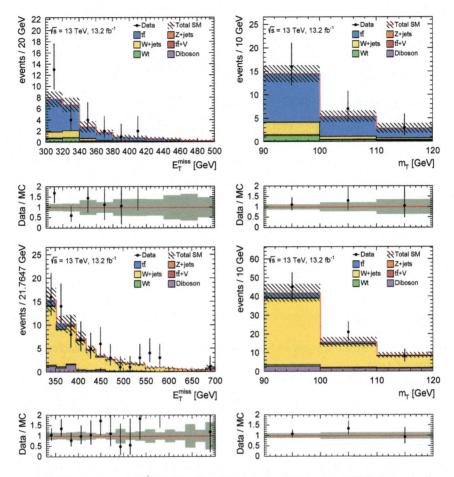

Fig. 9.18 Distributions of E_T^{miss} (left) and m_T (right) for the TVR (top) and WVR (bottom) selection associated with DM_high. Each background ($t\bar{t}$, W+jets, Wt, and $t\bar{t} + W/Z$) is normalised according to the result of the background-only fit to the control regions. Statistical and experimental systematic uncertainties are included in the error band

9.6.2.1 Non-canonical Validation Regions

A set of additional, "non-canonical" validation regions is defined. These VRs are not directly related to any SR, but try to probe particular phase spaces close to the SRs. Here, no normalisation factors are applied, since they do not coincide with any of the SRs targeted by the control regions. As mentioned above, dileptonic $t\bar{t}$ generally presents the largest background contribution after SR selection. In order to study this backgrounds in detail, a region selecting two leptons[7] is defined via:

---

[7]The m_T is constructed using the leading lepton.

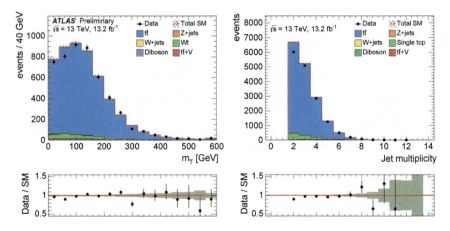

Fig. 9.19 Distributions of m_T and jet multiplicity for events passing the di-lepton validation region selection. The backgrounds are normalised to their cross section, no normalisation factors are applied and the uncertainty band includes statistical uncertainties

- dilepton trigger,
- quality cuts,
- exactly two signal leptons with different flavours ($e\mu$) to suppress contributions from Z+jets,
- the lepton pair has to be of opposite charge,
- ≥ 4 jets with $p_T > 25$ GeV.

Such events can only pass the signal selections if at least one extra jet is radiated from the initial or final state. Hence, it is particularly interesting to validate the modelling of the jet multiplicity distribution in this region. As can be seen in Fig. 9.19, it is found to be well-modelled for $n_{jets} \geq 4$. Also the m_T distribution agrees well between data and MC. A similar approach, using am_{T2} to enrich the sample in dileptonic $t\bar{t}$ with a missed lepton, comes to the same conclusion.

A similarly important component of the $t\bar{t}$ background features a lepton and a hadronic tau in the final state. A validation region for such processes is defined by selecting:

- Four jets with $p_T > 80, 50, 40, 25$ GeV,
- $E_T^{miss} > 200$ GeV,
- $m_T > 100$ GeV,
- At least one b-jet,
- One loose hadronic tau candidate and one signal lepton.

The corresponding m_T and am_{T2} distributions are shown in Fig. 9.20 and prove good data-MC agreement in the tails of the distributions which are relevant for the analysis.

The SRs cut tightly on the tail of the m_T distribution. Background events from $t\bar{t}$ or W+jets can only enter this region due to the finite resolution of the reconstructed

9.6 Background Estimation

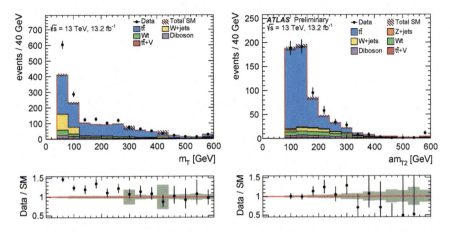

Fig. 9.20 Distributions of m_T and am_{T2} for events passing the $1\ell 1\tau$ validation region selection. The backgrounds are normalised to their cross sections, no normalisation factors are applied and the uncertainty band includes statistical uncertainties

m_T variable. In order to test the modelling in this regime while being orthogonal to the SRs, a b-veto is applied, enhancing the sample in W+jets events. The selection cuts are:

- Four jets with $p_T >$ 100, 80, 50, 25 GeV,
- $E_T^{miss} >$ 200 GeV,
- $m_T >$ 100 GeV,
- Exactly zero b-jets.

Figure 9.21 shows the distribution of the m_T and the E_T^{miss} distribution. The shape of the distributions in data is found to be reasonably well-modelled by MC, but the normalisation is off by at least 20%, consistent with the determined normalisation factors for W+jets.

Following a similar motivation, a validation region verifying the modelling in the am_{T2} tail is constructed in the following way:

- Four jets with $p_T >$ 80, 60, 60, 40 GeV,
- $E_T^{miss} >$ 200 GeV,
- $30 < m_T < 90$ GeV,
- $am_{T2} >$ 200 GeV,
- $H_{T,\text{sig}}^{miss} > 8$,
- Exactly 1 b-jet.

The data-MC agreement for distributions of am_{T2} and E_T^{miss} is presented in Fig. 9.22. The processes in this VR are found to be well-modelled with some discrepancy of 20–40% at very high am_{T2}, but with large uncertainties.

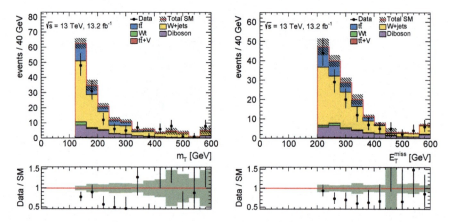

Fig. 9.21 Distributions of m_T and E_T^{miss} for events fulfilling the requirements of the W-tail validation region. The backgrounds are normalised to their cross sections, no normalisation factors are applied and the uncertainty band includes statistical uncertainties

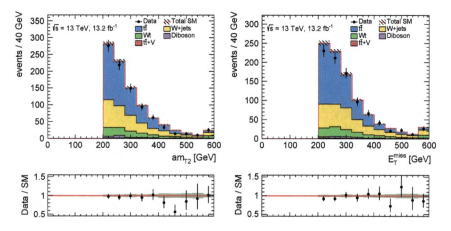

Fig. 9.22 Distributions of am_{T2} and E_T^{miss} for events fulfilling the requirements of the am_{T2}-tail validation region. The backgrounds are normalised to their cross sections, no normalisation factors are applied and the uncertainty band includes statistical uncertainties

9.7 Systematic Uncertainties

Since the MC background prediction is constrained from data in control regions, systematic uncertainties coming from experimental origins or theoretical aspects of prediction and modelling of backgrounds enter only in the extrapolation from the control region to other regions. The overall normalisation is not directly subject to systematic uncertainties. This extrapolation requires that the variables with cuts that differ between signal and control samples are very well understood. The sys-

9.7 Systematic Uncertainties

tematics are considered by the likelihood fits as nuisance parameters with Gaussian constraints.

9.7.1 Experimental Uncertainties

Uncertainties on the jet energy scale and resolution, on the treatment of the E_T^{miss} soft term and on the b-tagging efficiencies for b, c and light jets [46, 47] dominate in this analysis.

The transfer factor providing the extrapolation from control to SRs is affected by 4–15% (0–9%), depending on the SR, by the uncertainty on the jet energy scale (resolution). The uncertainty on the b-tagging amounts to a 0–6% effect on the extrapolation, the E_T^{miss} soft-term leads to 0–3% uncertainty.

Furthermore, uncertainties on lepton- and photon-related quantities (energy scales, resolutions, reconstruction and identification efficiencies, isolation) are considered and found to be negligible, as well as the uncertainty on the total integrated luminosity.

9.7.2 Theoretical Uncertainties

Since the extrapolation factor from control to SR, the *transfer factor*, is determined based on simulation, uncertainties on the modelling of the relevant background processes influence the result. The uncertainties on the $t\bar{t}$ and single-top modelling is estimated by varying the hadronisation and renormalisation scales and the strength of initial- and final state radiation that is assumed in the simulation [48]. Furthermore, different event generators, namely POWHEG-BOX with Herwig++ and MG5_aMC with Herwig++, have been tested. By interfacing POWHEG-BOX with different shower generators, namely PYTHIA 8 or Herwig++, the effect from fragmentation and hadronisation modelling is estimated. An additional uncertainty comes from interference between the $t\bar{t}$ and Wt processes at higher orders. This effect is estimated by comparing an inclusive $WWbb$ sample to the sum of the $t\bar{t}$ and Wt samples [48]. The observed difference is taken as uncertainty on the interference terms and is the dominant uncertainty on the Wt simulation. In summary, the extrapolation from TCRs and STCRs to SRs is affected by 17–32% for $t\bar{t}$ and by 14–68% for Wt events.

For the estimation of the theoretical uncertainties on the W+jets background, a different generator (MG5_aMC) is considered, renormalisation and factorisation scales are varied and different choices for the matrix element to parton shower matching and the resummation parameters are tested. While the generator dependence results in a 10–20% uncertainty, the effects from scale variations are found to be between 0 and 10%. An additional uncertainty needs to be considered due to the fact that SRs require at least one b-jet while the WCRs are constructed by applying a b-jet veto. The results presented in Ref. [49] are extrapolated to higher jet multiplicities and lead to an estimated uncertainty of around 40% for all SRs but bCbv, where it is found to be 20%.

As detailed above, a $t\bar{t}\gamma$ control region is used to normalise the $t\bar{t}Z$ background. This means that not only the extrapolation over kinematic quantities, as for the other control regions, but also the translation between the different physical processes is object to systematic uncertainties. A correction of 4% is applied to the $t\bar{t}\gamma$ sample to mitigate the differences of the details of the simulation used to generate the samples. The variation of renormalisation and factorisation scales affects $t\bar{t}Z$ and $t\bar{t}\gamma$ processes at leading order slightly differently, leading to an uncertainty of 10%. This is studied in detail by including NLO corrections in the form of k-factors that are applied. The k-factor is calculated for both processes using MG5_aMC or SHERPA plus OpenLoops for a nominal setup and with certain variations of the simulation parameters. The ratio of the resulting k-factors is then studied as a function of the boson p_T. The variation of the factorisation and renormalisation scales results in a 5% change of the k-factor ratio, assuming a different PDF set (NNPDF or CT14 [50]) leads to a difference of 2%. The dependence on the generator used introduces a 5% uncertainty. In total, an uncertainty on the extrapolation from the $t\bar{t}\gamma$ control region to the $t\bar{t}Z$ component of the SR of 12% is found.

The diboson estimate from MC is subject to uncertainties on the cross section of around 6%. Together with the results from varying the renormalisation and factorisation scales during simulation, a total theoretical uncertainty of 20–30% is assumed.

Variations of renormalisation and factorisation scales as well as PDF sets are taken into account for the SUSY signal samples, resulting in an uncertainty of 13–23%. Only the effect on the signal acceptance is taken as uncertainty when the factorisation and renormalisation scales for the leading-order DM signals are evaluated, resulting in a 5% uncertainty.

9.8 Results

The measured event yield in data together with the estimated background prediction from the likelihood fit to the CRs is presented in Table 9.6 for the SRs and in Fig. 9.23 for SRs and VRs. Note that neither the SRs nor the validation regions are disjoint. The normalisation factors for the different backgrounds, as obtained from the background-only fit, are also listed in Table 9.6. The background prediction is considered conservative since any signal contamination in the control regions is attributed to background processes and thus possibly yields to an overestimation of the background in the SR. The compatibility of the observed results with the background-only hypothesis is tested by the likelihood ratio, using the CL_s prescription [46]. The resulting probabilities of the background-only hypothesis appear as *p-values* p_0 in Table 9.6. Furthermore, the fit is repeated treating the signal strength as a floating parameter to determine the limit on the number of events from new physics in the SRs, quoted as N^{limit}_{non-SM} in Table 9.6.

9.8 Results

Table 9.6 Data events and expected background yields and their uncertainties as predicted by the background-only fits in the SRs. The background normalisation factors obtained by the background-only fit (NF), and the probabilities (p_0) of the background-only hypothesis given the observed result are shown as well

Signal region	SR1	tN_high	bC2x_diag	bC2x_med	bCbv	DM_low	DM_high
Observed	37	5	37	14	7	35	21
Total background	24 ± 3	3.8 ± 0.8	22 ± 3	13 ± 2	7.4 ± 1.8	17 ± 2	15 ± 2
$t\bar{t}$	8.4 ± 1.9	0.60 ± 0.27	6.5 ± 1.5	4.3 ± 1.0	0.26 ± 0.18	4.2 ± 1.3	3.3 ± 0.8
W+jets	2.5 ± 1.1	0.15 ± 0.38	1.2 ± 0.5	0.63 ± 0.29	5.4 ± 1.8	3.1 ± 1.5	3.4 ± 1.4
Single top	3.1 ± 1.5	0.57 ± 0.44	5.3 ± 1.8	5.1 ± 1.6	0.24 ± 0.23	1.9 ± 0.9	1.3 ± 0.8
$t\bar{t} + V$	7.9 ± 1.6	1.6 ± 0.4	8.3 ± 1.7	2.7 ± 0.7	0.12 ± 0.03	6.4 ± 1.4	5.5 ± 1.1
Diboson	1.2 ± 0.4	0.61 ± 0.26	0.45 ± 0.17	0.42 ± 0.20	1.1 ± 0.4	1.5 ± 0.6	1.4 ± 0.5
Z+jets	0.59 ± 0.54	0.03 ± 0.03	0.32 ± 0.29	0.08 ± 0.08	0.22 ± 0.20	0.16 ± 0.14	0.47 ± 0.44
$t\bar{t}$ NF	1.03 ± 0.07	1.06 ± 0.15	0.89 ± 0.10	0.95 ± 0.12	0.73 ± 0.22	0.90 ± 0.17	1.01 ± 0.13
W+jets NF	0.76 ± 0.08	0.78 ± 0.08	0.87 ± 0.07	0.85 ± 0.06	0.97 ± 0.12	0.94 ± 0.13	0.91 ± 0.07
Single top NF	1.07 ± 0.30	1.30 ± 0.45	1.26 ± 0.31	0.97 ± 0.28	–	1.36 ± 0.36	1.02 ± 0.32
$t\bar{t} + W/Z$ NF	1.43 ± 0.21	1.39 ± 0.22	1.40 ± 0.21	1.30 ± 0.23	–	1.47 ± 0.22	1.42 ± 0.21
p_0 (σ)	0.012 (2.2)	0.26 (0.6)	0.004 (2.6)	0.40 (0.3)	0.50 (0)	0.0004 (3.3)	0.09 (1.3)
$N_{\text{non-SM}}^{\text{limit}}$ exp. (95% CL)	$12.9^{+5.5}_{-3.8}$	$5.5^{+2.8}_{-1.1}$	$12.4^{+5.4}_{-3.7}$	$9.0^{+4.2}_{-2.7}$	$7.3^{+3.5}_{-2.2}$	$11.5^{+5.0}_{-3.4}$	$9.9^{+4.6}_{-2.9}$
$N_{\text{non-SM}}^{\text{limit}}$ obs. (95% CL)	26.0	7.2	27.5	9.9	7.2	28.3	15.6

In three SRs deviations larger than two standard deviations from the background prediction are observed. The largest excess with a local significance of 3.3σ is seen in the DM_low SR, the excess in SR1 amounts to 2.2σ (local) and bC2x_diag sees an excess of 2.6σ (local).

As mentioned, there is a potential overlap between signal and background events also in these SRs. As a cross-check, exclusive SRs were defined by excluding events that are also selected by another of these three SRs. When done so, the excess in SR1 is reduced, while the deviations in DM_low and bC2x_diag persist.

The distributions of E_T^{miss} and m_T in the DM SRs and the SRs seeing an excess (SR1 and bC2x_diag) are presented in Figs. 9.24 and 9.25. In all three deviating SRs the excess tends to favour low values of m_T and of E_T^{miss}.

With the observation of an excess over 3σ many additional cross-checks were performed to scrutinise the analysis procedure and its results. A subset is presented in Appendix D, along with additional distributions at preselection level and in the DM control, validation and SRs in Appendix C. These checks did not reveal any particular problem with the analysis and its methods that could cause the observed discrepancies between background expectation and observed data events.

204 9 Search for New Physics in Events with Missing Energy and Top Quarks

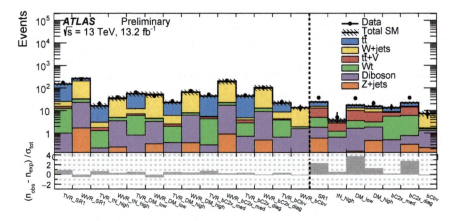

Fig. 9.23 Summary of the observed event yields in data (n_{obs}) compared to the predicted background (n_{exp}) in the VR and SRs. The bottom panel shows the significance of the deviation between data and background expectation. The significance considers the total uncertainty (σ_{tot})

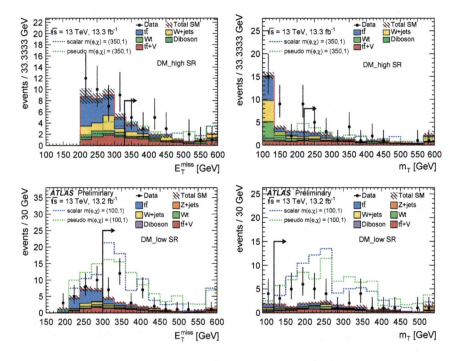

Fig. 9.24 The E_T^{miss} (left) and m_T (right) distributions in DM_high (top) and DM_low (bottom). In each plot, the full event selection in the corresponding signal region is applied, except for the requirement (indicated by an arrow) that is imposed on the variable being plotted. The predicted backgrounds are scaled with the normalisation factors documented in Table 9.6. The uncertainty band includes statistical uncertainties. The last bin contains the overflow. Benchmark signal models where a common coupling $g = g_q = g_\chi = 3.5$ is assumed are overlaid for comparison

9.9 Interpretation of the Results

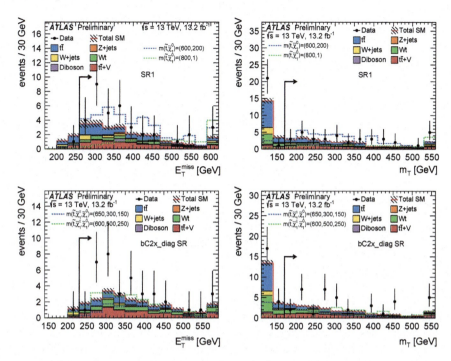

Fig. 9.25 The E_T^{miss} (left) and m_T (right) distributions in SR1 (top) and bC2x_diag (bottom). In each plot, the full event selection in the corresponding signal region is applied, except for the requirement (indicated by an arrow) that is imposed on the variable being plotted. The predicted backgrounds are scaled with the normalisation factors documented in Table 9.6. The uncertainty band includes statistical uncertainties. The last bin contains the overflow. Benchmark signal models are overlaid for comparison

9.9 Interpretation of the Results

The measured events in data in view of the estimated background expectation can be used to set limits on the various signal scenarios.

The exclusion fit tests the signal plus background hypothesis. In the exclusion fit, all control regions and the SR are used in the fit. The signal contribution in all regions is taken into account as predicted by the signal model under study. The exclusion test result is a CL_s value, which represents the probability for the observation being compatible with the signal plus background hypothesis.

Two types of CL_s values are presented: the *expected* CL_s value is obtained by setting the data in the SRs equal to the total fitted background prediction, and provides an estimate of the expected sensitivity of the analysis. The *observed* CL_s value is obtained when the observed data is considered in the SRs. For CL_s values below 0.05 the given signal model is excluded at 95% confidence level.

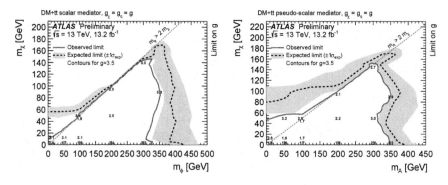

Fig. 9.26 The observed upper limits on the combined coupling $g_{q\chi}$ for each signal grid point is presented as numbers in the plane of m_ϕ (m_A) versus m_χ for the Simplified Model describing Dark Matter production in association with top quarks for a scalar mediator (left) and pseudo-scalar mediator (right). Expected (dashed) and observed (solid) 95% CL exclusion contours assume a maximal coupling of $g = 3.5$

Under a certain signal hypothesis, possible signal contributions in the control regions are taken into account in the fit, as well as all uncertainties, except for the theoretical ones on the signal cross section.[8] The exclusion limits are derived at 95% CL. For each signal point the SR providing the lowest CL_s value is chosen. The exclusion contours are determined via an interpolation on the calculated CL_s values.

9.9.1 Limits on Dark Matter Models

The results are interpreted in terms of the Simplified Model of DM pair production, assuming scalar or pseudo-scalar mediators. The limits are presented in Fig. 9.26, as upper bounds on the combined coupling $g = g_q = g_\chi$, and as exclusion contours under the assumption of a maximal coupling of $g = 3.5$. These contours extends up to mediator masses of 350 GeV for a light DM particle.

In discussions after the presentation of these results, it was brought up that for the excluded regions at high mediator mass the decay width of the mediator starts to become sufficiently large to maybe alter the kinematic behaviour of the signal, which would prevent to quote a limiting coupling strength purely based on the signal cross section, as was done here. A compromise would be to define exclusion contours in the future for lower couplings of e.g. $g = 2$. Ideally, a natural coupling of one would be assumed, but in the present case the sensitivity of the analysis is still beyond an exclusion for this coupling choice.

---

[8] For SUSY limits these are quoted as a band around the observed limit.

9.9 Interpretation of the Results

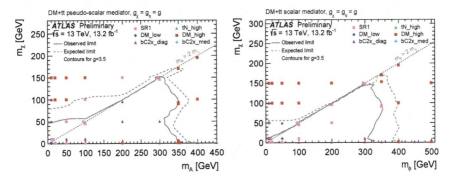

Fig. 9.27 Illustration of the best expected signal region per signal grid point in the plane of m_ϕ (m_A) versus m_χ for Dark Matter production associated production with top quarks for a scalar mediator (left) and pseudo-scalar mediator (right)

For many DM signal points another SR is found to perform slightly better than DM_low or DM_high, which is shown in Fig. 9.27.[9] However, the expected limits obtained only from the DM regions, presented in Fig. 9.28, are comparable to the one where the best performing SR is chosen.[10] This is a manifestation of the fact that all of the more inclusive SRs (DM, SR1, bC2x_diag) perform similarly well in constraining the DM signals in the probed phase-space. This similarity could well be broken when considering more (or less) data.

9.9.2 Limits on Direct Stop Production

The analysis is able to extend the excluded stop mass range up to 830 GeV for a very light neutralino $\tilde{\chi}_1^0$ in the tN scenario, where $\mathcal{BR}(\tilde{t}_1 \to t + \tilde{\chi}_1^0) = 100\%$ is assumed. The stop-neutralino mass plane with the expected and observed exclusion limit contours is shown in Fig. 9.29. Assuming a bC scenario with $\mathcal{BR}(\tilde{t}_1 \to b + \tilde{\chi}_1^\pm) = 100\%$ and fixing $m_{\tilde{\chi}_1^\pm} = 2m_{\tilde{\chi}_1^0}$, stop masses up to 750 GeV are excluded with a 150 GeV neutralino, as can be seen in Fig. 9.30, where expected and observed limit contours are shown in the stop-neutralino mass plane. In the case of a small mass gap between stop and chargino, $m_{\tilde{\chi}_1^\pm} = m_{\tilde{t}_1} - 10$ GeV, the exclusion in terms of the stop mass extends up to 750 GeV for a very light neutralino, as shown in Fig. 9.30.

---

[9]This also holds vice versa: for some tN and bC signal points, the DM regions were found to perform best.

[10]For DM_low, this is only the case for low mediator masses, for which this region was optimised for.

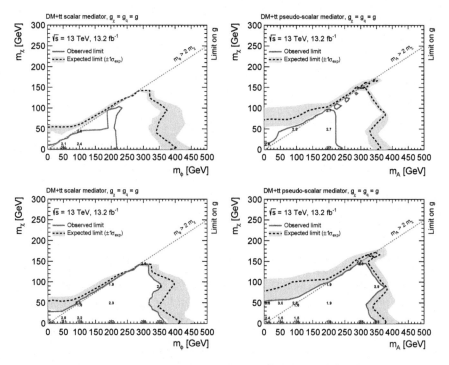

Fig. 9.28 The observed 95% CL upper limit on the couplings in the plane of m_ϕ (m_A) versus m_χ for DM associated production with top quarks for a scalar (pseudo-scalar) mediator. The results from DM_low is shown on top and for DM_high on the bottom, for scalar (left) and pseudo-scalar (right) mediators. The numbers on the plot show the value of the excluded coupling for the corresponding points on the signal grid

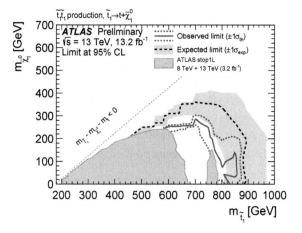

Fig. 9.29 Expected (black dashed) and observed (red solid) exclusion contours at 95% CL in the plane of $m_{\tilde{t}_1}$ versus $m_{\tilde{\chi}_1^0}$. Direct stop pair production is assumed, where the $\tilde{t}_1 \to t + \tilde{\chi}_1^0$ decay is fixed to have a branching ratio of 100%. While the limits from earlier analyses [8, 36], indicated by the grey shaded area, assume mostly-right-handed stops, the present results consider unpolarised stops

9.9 Interpretation of the Results

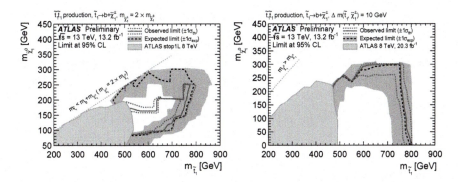

Fig. 9.30 Expected (black dashed) and observed (red solid) exclusion contours at 95% CL in the plane of $m_{\tilde{t}_1}$ versus $m_{\tilde{\chi}_1^0}$. Direct stop pair production is assumed, where the $\tilde{t}_1 \to b + \tilde{\chi}_1^\pm$ decay is fixed to have a branching ratio of 100%. A chargino mass of twice the neutralino mass (left) or a small mass difference with respect to the stop, $m_{\tilde{\chi}_1^\pm} = m_{\tilde{t}_1} - 10$ GeV (right) are considered. While the limits from earlier analyses [8, 36], indicated by the grey shaded area, assume mostly-left-handed stops, the present results consider unpolarised stops

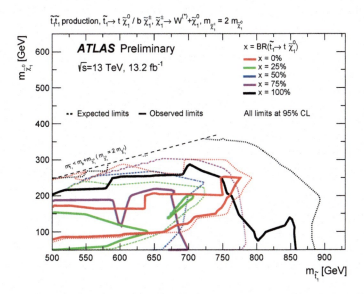

Fig. 9.31 Expected (dashed) and observed (solid) 95% excluded regions in the plane of $m_{\tilde{t}_1}$ versus $m_{\tilde{\chi}_1^0}$ for direct stop pair production for different assumptions of $x = \mathcal{BR}(\tilde{t}_1 \to t + \tilde{\chi}_1^0) = 1 - \mathcal{BR}(\tilde{t}_1 \to b + \tilde{\chi}_1^\pm)$ where the chargino mass is assumed to be twice the neutralino mass, and x is scanned from 0 to 100% in steps of 25%. No points can be excluded in data for the $x = 50\%$ scenario

The production of unpolarised stops is assumed here. The resulting limits are expected to be slightly weaker than if left-handed stops were assumed for the tN decay or if right-handed stops were assumed for the bC scenarios.

In a so-called *mixed scenario* both decays $\tilde{t}_1 \to t + \tilde{\chi}_1^0$ and $\tilde{t}_1 \to b + \tilde{\chi}_1^\pm$, under the assumption of $m_{\tilde{\chi}_1^\pm} = 2m_{\tilde{\chi}_1^0}$, are allowed. Branching ratios of 0, 25, 50, 75 and 100 % for $\tilde{t}_1 \to t + \tilde{\chi}_1^0$ are considered while $\mathcal{BR}\,(\tilde{t}_1 \to b + \tilde{\chi}_1^\pm) = 100\% - \mathcal{BR}\,(\tilde{t}_1 \to t + \tilde{\chi}_1^0)$. The resulting exclusion contours are presented in Fig. 9.31.

9.10 Conclusions

A dataset of $13.2\,\text{fb}^{-1}$ of $\sqrt{s} = 13\,\text{TeV}$ proton-proton collisions, recorded by ATLAS, has been analysed in search for new physics in events with E_T^{miss} and top quarks for final states with one lepton. Dedicated signal regions were designed to target stop decays to $\tilde{t}_1 \to t + \tilde{\chi}_1^0$ and $\tilde{t}_1 \to b + \tilde{\chi}_1^\pm$, as well as for scalar and pseudo-scalar mediated DM production. The dominant sources of background are constrained in signal-region-specific control regions enriched in $t\bar{t}$, W+jets, Wt or $t\bar{t}\gamma$ (for the estimate of $t\bar{t}Z$). This strategy allows to significantly reduce the impact of systematic uncertainties on the result. The largest experimental source is represented by the jet energy scale uncertainty and amounts to 4–15%, depending on the signal region.

An excess of 3.3σ of data events over the background prediction is found in the DM_low signal region. Also the SR1 and bC2x_diag signal regions see more data events than expected, with a deviation above 2σ. In all other signal regions, as well as in the validation regions constructed to monitor the background modelling, agreement between the prediction and the measurement is observed.

The observations are translated into limits on the targeted signal models. In both the $\tilde{t}_1 \to t + \tilde{\chi}_1^0$ and the $\tilde{t}_1 \to b + \tilde{\chi}_1^\pm$ scenario, the exclusion bounds reach beyond precedent results.

For the first time within ATLAS, a Simplified Model is used to present the bounds on DM pair production in association with top quarks. The results are presented as upper limits at 95% CL on the combined coupling strength for each grid point as well as an exclusion contour in the M_{med}–m_χ mass plane, assuming a coupling of 3.5.

During preparation and after presentation of these results, more collision data was recorded. The analysis was then repeated on a dataset of about twice the integrated luminosity as considered for the presented results. None of the excesses could be confirmed. Nevertheless, this powerful analysis is further improved and will likely be able to continue to provide interesting results on the top squark production and the production of Dark Matter with top quarks.

References

1. D. Abercrombie et al., Dark matter benchmark models for early LHC Run-2 searches: report of the ATLAS/CMS dark matter forum, arXiv:1507.00966 [hep-ex]
2. M.R. Buckley, D. Feld, D. Goncalves, Scalar simplified models for dark matter. Phys. Rev. **D91**, 015017 (2015). https://doi.org/10.1103/PhysRevD.91.015017, arXiv:1410.6497 [hep-ph]
3. U. Haisch, E. Re, Simplified dark matter top-quark interactions at the LHC. JHEP **06**, 078 (2015). https://doi.org/10.1007/JHEP06(2015)078, arXiv:1503.00691 [hep-ph]
4. ATLAS Collaboration, Search for the supersymmetric partner of the top quark in the Jets+Emiss final state at $\sqrt{s} = 13$ TeV. Technical Report, ATLAS-CONF-2016-077, CERN, Geneva, Aug 2016, https://cds.cern.ch/record/2206250
5. ATLAS Collaboration, Search for direct top squark pair production and dark matter production in final states with two leptons in $\sqrt{s} = 13$ TeV pp collisions using 13.3 fb^{-1} of ATLAS data. Technical Report, ATLAS-CONF-2016-076, CERN, Geneva, Aug 2016, https://cds.cern.ch/record/2206249
6. ATLAS Collaboration, G. Aad et al., Search for dark matter in events with heavy quarks and missing transverse momentum in pp collisions with the ATLAS detector. Eur. Phys. J. **C75**(2), 92 (2015).https://doi.org/10.1140/epjc/s10052-015-3306-z, arXiv:1410.4031 [hep-ex]
7. CMS Collaboration, Search for the production of dark matter in association with top-quark pairs in the single-lepton final state in proton-proton collisions at $\sqrt{s} = 8$ TeV. J. High Energy Phys. **06**(121), 35 pp. (2015). CMS-B2G-14-004. CERN-PH-EP-2015-087, http://cds.cern.ch/record/2008743, arXiv:1504.03198
8. ATLAS Collaboration, M. Aaboud et al., Search for top squarks in final states with one isolated lepton, jets, and missing transverse momentum in $\sqrt{s} = 13$ TeV pp collisions with the ATLAS detector. Phys. Rev. **D94**(5), 052009 (2016). https://doi.org/10.1103/PhysRevD.94.052009, arXiv:1606.03903 [hep-ex]
9. ATLAS Collaboration, G. Aad et al., Improved luminosity determination in pp collisions at $\sqrt{s} = 7$ TeV using the ATLAS detector at the LHC. Eur. Phys. J. **C73**(8), 2518 (2013). https://doi.org/10.1140/epjc/s10052-013-2518-3, arXiv:1302.4393 [hep-ex]
10. ATLAS Collaboration, M. Aaboud et al., Luminosity determination in pp collisions at $\sqrt{s} = 8$ TeV using the ATLAS detector at the LHC, arXiv:1608.03953 [hep-ex]
11. S. Alioli, P. Nason, C. Oleari, E. Re, A general framework for implementing NLO calculations in shower Monte Carlo programs: the POWHEG BOX. JHEP **1006**, 043 (2010). https://doi.org/10.1007/JHEP06(2010)043, arXiv:1002.2581 [hep-ph]
12. T. Sjostrand, S. Mrenna, P.Z. Skands, PYTHIA 6.4 physics and manual. JHEP **05**, 026 (2006). https://doi.org/10.1088/1126-6708/2006/05/026, arXiv:hep-ph/0603175 [hep-ph]
13. T. Gleisberg, S. Hoeche, F. Krauss, M. Schonherr, S. Schumann, F. Siegert, J. Winter, Event generation with SHERPA 1.1. JHEP **02**, 007 (2009). https://doi.org/10.1088/1126-6708/2009/02/007, arXiv:0811.4622 [hep-ph]
14. J. Alwall, R. Frederix, S. Frixione, V. Hirschi, F. Maltoni, O. Mattelaer, H.S. Shao, T. Stelzer, P. Torrielli, M. Zaro, The automated computation of tree-level and next-to-leading order differential cross sections, and their matching to parton shower simulations. JHEP **1407**, 079 (2014). https://doi.org/10.1007/JHEP07(2014)079, arXiv:1405.0301 [hep-ph]
15. T. Sjöstrand, S. Mrenna, P.Z. Skands, A brief introduction to PYTHIA 8.1. Comput. Phys. Commun. **178**, 852 (2008). https://doi.org/10.1016/j.cpc.2008.01.036, arXiv:0710.3820 [hep-ph]
16. K. Melnikov, M. Schulze, A. Scharf, QCD corrections to top quark pair production in association with a photon at hadron colliders. Phys. Rev. **D83**, 074013 (2011). https://doi.org/10.1103/PhysRevD.83.074013, arXiv:1102.1967 [hep-ph]
17. S.P. Martin, A supersymmetry primer, arXiv:hep-ph/9709356 [hep-ph]
18. GEANT4 Collaboration, S. Agostinelli, et al., (GEANT4 Collaboration), GEANT4: a simulation toolkit. Nucl. Instrum. Meth. **A506**, 250–303 (2003). https://doi.org/10.1016/S0168-9002(03)01368-8

19. ATLAS Collaboration, M. Beckingham, M. Duehrssen, E. Schmidt, M. Shapiro, M. Venturi, J. Virzi, I. Vivarelli, M. Werner, S. Yamamoto, T. Yamanaka, The simulation principle and performance of the ATLAS fast calorimeter simulation FastCaloSim. Technical Report, ATL-PHYS-PUB-2010-013, CERN, Geneva, Oct 2010, http://cds.cern.ch/record/1300517
20. ATLAS Collaboration Collaboration, Electron and photon energy calibration with the ATLAS detector using data collected in 2015 at $\sqrt{s} = 13$ TeV. Technical Report, ATL-PHYS-PUB-2016-015, CERN, Geneva, Aug 2016, http://cds.cern.ch/record/2203514
21. ATLAS Collaboration, G. Aad et al., Muon reconstruction performance of the ATLAS detector in proton-proton collision data at $\sqrt{s} = 13$ TeV. Eur. Phys. J. **C76**(5), 292 (2016). https://doi.org/10.1140/epjc/s10052-016-4120-y, arXiv:1603.05598 [hep-ex]
22. ATLAS Collaboration, Electron identification measurements in ATLAS using $\sqrt{s} = 13$ TeV data with 50 ns bunch spacing. Technical Report, ATL-PHYS-PUB-2015-041, CERN, Geneva, Sep 2015, http://cds.cern.ch/record/2048202
23. ATLAS Collaboration, Electron efficiency measurements with the ATLAS detector using the 2015 LHC proton-proton collision data. Technical Report, ATLAS-CONF-2016-024, CERN, Geneva, Jun 2016, http://cds.cern.ch/record/2157687
24. ATLAS Collaboration, Expected photon performance in the ATLAS experiment. Technical Report, ATL-PHYS-PUB-2011-007, CERN, Geneva, Apr 2011, http://cds.cern.ch/record/1345329
25. ATLAS Collaboration, G. Aad et al., Electron reconstruction and identification efficiency measurements with the ATLAS detector using the 2011 LHC proton-proton collision data. Eur. Phys. J. **C74**(7), 2941 (2014). https://doi.org/10.1140/epjc/s10052-014-2941-0, arXiv:1404.2240 [hep-ex]
26. ATLAS Collaboration, G. Aad et al., Characterisation and mitigation of beam-induced backgrounds observed in the ATLAS detector during the 2011 proton-proton run. JINST **8**, P07004 (2013). https://doi.org/10.1088/1748-0221/8/07/P07004, arXiv:1303.0223 [hep-ex]
27. ATLAS Collaboration, Selection of jets produced in 13 TeV proton-proton collisions with the ATLAS detector. Technical Report, ATLAS-CONF-2015-029, CERN, Geneva, Jul 2015, http://cds.cern.ch/record/2037702
28. ATLAS Collaboration, Optimisation of the ATLAS b-tagging performance for the 2016 LHC Run. Technical Report, ATL-PHYS-PUB-2016-012, CERN, Geneva, Jun 2016, https://cds.cern.ch/record/2160731
29. ATLAS Collaboration, G. Aad et al., Performance of b-Jet Identification in the ATLAS Experiment. JINST **11**(04), P04008 (2016). https://doi.org/10.1088/1748-0221/11/04/P04008, arXiv:1512.01094 [hep-ex]
30. ATLAS Collaboration, Commissioning of the reconstruction of hadronic tau lepton decays in ATLAS using pp collisions at $\sqrt{s} = 13$ TeV. Technical Report, ATL-PHYS-PUB-2015-025, CERN, Geneva, Jul 2015, http://cds.cern.ch/record/2037716
31. ATLAS Collaboration, Reconstruction, energy calibration, and identification of hadronically decaying tau leptons in the ATLAS experiment for Run-2 of the LHC. Technical Report, ATL-PHYS-PUB-2015-045, CERN, Geneva, Nov 2015, http://cds.cern.ch/record/2064383
32. ATLAS Collaboration, Performance of missing transverse momentum reconstruction for the ATLAS detector in the first proton-proton collisions at at $\sqrt{s} = 13$ TeV. Technical Report, ATL-PHYS-PUB-2015-027, CERN, Geneva, Jul 2015, http://cds.cern.ch/record/2037904
33. ATLAS Collaboration, Expected performance of missing transverse momentum reconstruction for the ATLAS detector at $\sqrt{s} = 13$ TeV. Technical Report, ATL-PHYS-PUB-2015-023, CERN, Geneva, Jul 2015, http://cds.cern.ch/record/2037700
34. ATLAS Collaboration, Search for direct pair production of the top squark in all-hadronic final states in proton-proton collisions at $\sqrt{s} = 8$ TeV with the ATLAS detector. J. High Energy Phys. **09**(015) 57 pp. (2014). CERN-PH-EP-2014-112, http://cds.cern.ch/record/1706342, arXiv:1406.1122
35. M. Cacciari, G.P. Salam, G. Soyez, The catchment area of jets. JHEP **0804**, 005 (2008). https://doi.org/10.1088/1126-6708/2008/04/005, arXiv:0802.1188 [hep-ph]

References

36. ATLAS Collaboration, Search for top squark pair production in final states with one isolated lepton, jets, and missing transverse momentum in $\sqrt{s} = 8$ TeV pp collisions with the ATLAS detector. J. High Energy Phys. **11**(118), 94 pp. (2014). CERN-PH-EP-2014-143, http://cds.cern.ch/record/1714148, arXiv:1407.0583
37. M.L. Graesser, J. Shelton, Hunting mixed top squark decays. Phys. Rev. Lett. **111**(12), 121802 (2013). https://doi.org/10.1103/PhysRevLett.111.121802, arXiv:1212.4495 [hep-ph]
38. C.G. Lester, D.J. Summers, Measuring masses of semi-invisibly decaying particles pair produced at hadron colliders. Phys. Lett. **B463**, 99–103 (1999). https://doi.org/10.1016/S0370-2693(99)00945-4, arXiv:hep-ph/9906349 [hep-ph]
39. A.J. Barr, B. Gripaios, C.G. Lester, Transverse masses and kinematic constraints: from the boundary to the crease. JHEP **0911**, 096 (2009). https://doi.org/10.1088/1126-6708/2009/11/096, arXiv:0908.3779 [hep-ph]
40. P. Konar, K. Kong, K.T. Matchev, M. Park, Dark matter particle spectroscopy at the LHC: generalizing M(T2) to asymmetric event topologies. JHEP **1004**, 086 (2010). https://doi.org/10.1007/JHEP04(2010)086, arXiv:0911.4126 [hep-ph]
41. Y. Bai, H.-C. Cheng, J. Gallicchio, J. Gu, Stop the top background of the stop search. JHEP **1207**, 110 (2012). https://doi.org/10.1007/JHEP07(2012)110, arXiv:1203.4813 [hep-ph]
42. C.G. Lester, B. Nachman, Bisection-based asymmetric M_{T2} computation: a higher precision calculator than existing symmetric methods. JHEP **1503**, 100 (2015). https://doi.org/10.1007/JHEP03(2015)100, arXiv:1411.4312 [hep-ph]
43. B. Nachman, C.G. Lester, Significance variables. Phys. Rev. **D88**(7), 075013 (2013). https://doi.org/10.1103/PhysRevD.88.075013, arXiv:1303.7009 [hep-ph]
44. W. Verkerke et al., The roofit toolkit for data modeling, http://roofit.sourceforge.net or with recent versions of the root framework, http://root.cern.ch
45. M. Baak, G.J. Besjes, D. Côte, A. Koutsman, J. Lorenz, D. Short, HistFitter software framework for statistical data analysis. Eur. Phys. J. **C75**, 153 (2015). https://doi.org/10.1140/epjc/s10052-015-3327-7, arXiv:1410.1280 [hep-ex]
46. ATLAS Collaboration, Calibration of b-tagging using dileptonic top pair events in a combinatorial likelihood approach with the ATLAS experiment. Technical Report, ATLAS-CONF-2014-004, CERN, Geneva, Feb 2014, http://cds.cern.ch/record/1664335
47. ATLAS Collaboration, Calibration of the performance of b-tagging for c and light-flavour jets in the 2012 ATLAS data. Technical Report, ATLAS-CONF-2014-046, CERN, Geneva, Jul 2014, http://cds.cern.ch/record/1741020
48. ATLAS Collaboration, Simulation of top quark production for the ATLAS experiment at $\sqrt{s} = 13$ TeV. Technical Report, ATL-PHYS-PUB-2016-004, CERN, Geneva, Jan 2016, http://cds.cern.ch/record/2120417
49. ATLAS Collaboration, Measurement of the cross-section for W boson production in association with b-jets in pp collisions at $\sqrt{s} = 7$ TeV with the ATLAS detector. J. High Energy Phys. **06**(084), 53 pp. (2013). CERN-PH-EP-2012-357, http://cds.cern.ch/record/1516003, arXiv:1302.2929
50. S. Dulat, T.-J. Hou, J. Gao, M. Guzzi, J. Huston, P. Nadolsky, J. Pumplin, C. Schmidt, D. Stump, C.P. Yuan, New parton distribution functions from a global analysis of quantum chromodynamics. Phys. Rev. **D93**(3), 033006 (2016). https://doi.org/10.1103/PhysRevD.93.033006, arXiv:1506.07443 [hep-ph]
51. A.L. Read, Presentation of search results: The CL(s) technique. J. Phys. **G28**, 2693–2704 (2002). https://doi.org/10.1088/0954-3899/28/10/313

Chapter 10
Conclusions

From the discovery of the electron to the observation of the Higgs boson, particle physics has come a long way, both paved by experimental and theoretical advancements, culminating in the formulation and confirmation of the incredibly successful Standard Model of particle physics. As discussed in this thesis, albeit its success and the extremely precise experimental confirmations of its predictions the Standard Model cannot be the end: too many questions and puzzles are left unanswered. A plethora of interesting new-physics models is proposed to address one or several of these problems, and is scrutinised by LHC experiments. One well-motivated extension is Supersymmetry, which would solve many questions at once. Another strong motivation to search for physics beyond the Standard Model is the overwhelming experimental evidence from astrophysics for a non-luminous matter component named Dark Matter, for which the Standard Model cannot offer any candidate particle.

This thesis presented several aspects of searches for Dark Matter at the LHC. The presented work revealed that in a large fraction of collisions at the LHC the momentum transfer is well above the bound set on the cut-off scale of an effective field theory, rendering such interpretations inadequate. From the detailed discussion of the problems with an effective field theoretic interpretation of LHC searches, the need for a change of the way in which results are presented was motivated. In order to quantify the impact of this non-validity a rescaling procedure was suggested. The ATLAS search for new physics in monojet-like events adapted this rescaling procedure in the interpretation of its results. The thesis further presented this analysis of 20.3 fb^{-1} of 8 TeV pp collision data targeting final states with an energetic jet and large missing transverse energy. An optimisation study regarding signals of Dark Matter showed that it is beneficial to not explicitly veto additional jets in the event but rather ensure a monojet-like event selection via topological cuts. Furthermore, a veto on isolated tracks was developed which allowed to reduce the electroweak backgrounds significantly, especially those containing tau leptons. The background estimation technique concentrates on the large irreducible component from $Z(\rightarrow \nu\bar{\nu})$.

By combining the information from several one- and two-lepton control regions, the total uncertainty on the background prediction in the signal regions was achieved to a precision of 2.7–14%. No deviation from the estimated Standard Model background prediction was observed in data. Apart from the above-mentioned rescaled limits on the effective-theory cut-off scale, a so-called Simplified Model was considered in the interpretation, although the range of parameters was limited. A detailed reinterpretation of this and two other ATLAS searches for Dark Matter was presented subsequently. Three different Simplified Models were considered and a large range of parameters was tested. The study revealed that the searches could profit from a detailed optimisation for Simplified Models, especially in the regime of smaller missing transverse energies. Furthermore, the results included a comparison to bounds from direct searches for Dark Matter, which, within a Simplified Model, can be considered more reliable than within an effective field theory, as done earlier.

Simplified Models of Dark Matter production which feature a (pseudo-) scalar mediator would favour final states with heavy quarks: if Higgs-like Yukawa couplings and minimal flavour violation are assumed, the coupling is proportional to the mass of the interacting quark. A search for new physics in events with large missing transverse energy and a top quark pair in the one-lepton channel was presented. It considered 13.2 fb^{-1} of 13 TeV pp collision data. The main backgrounds from processes involving tops of W bosons are efficiently reduced by the use of specifically designed kinematic variables. The remaining contributions to the event yield after the signal selection is estimated in dedicated background-enriched control regions, which allows for a cancellation of many systematic effects. In three of the signal regions an excess of data over the background prediction was observed, with a local significance of up to 3.3σ. It is found towards lower values of transverse mass and missing transverse energy. Many additional tests could not assign the excess to an experimental artefact. The results were interpreted in terms of the production of supersymmetric partners of the top quark. In all considered signal scenarios, previous exclusions were extended. Also the parameters of a scalar Simplified Model of Dark Matter production could be constrained by the results. For the first time, such a model was used in the Dark Matter interpretation of a search with heavy quarks and missing transverse energy. Although the excess could not be confirmed when repeating the analysis on a larger dataset (during the preparation of this thesis), the analysis has a strong potential to discover new physics or strongly constrain the parameters of models predicting new particles in such a final state. The second run of the LHC has just started and much more interesting data is expected in the coming years. It is a unique situation to witness such an enhancement in the achieved centre-of-mass energy, opening doors for learning more about the world.

Within the collider experiments, the presented work showed different ways of how to profit from the precious data the LHC delivers in terms of Dark Matter. The *Monojet Analysis* analysis, traditionally the most important general Dark Matter search, was improved by generalising the considered final state which increased the signal acceptance significantly, and started the transition from effective field theories to theoretically more sane Simplified Models for interpretation. The latter change led to a wider paradigm shift concerning general Dark Matter searches: although

10 Conclusions

the LHC has potential to constrain Dark Matter production in mono-X searches–and does so–, constraints on the mediator itself via resonance searches are often more powerful. With the rise of Simplified Models, such resonance searches became more and more included in the Dark Matter discussions. For future Dark Matter searches at the LHC it might be beneficial to focus on the strengths of an LHC Dark Matter programme and not compete with direct mediator searches but rather focus on–often experimentally challenging–regimes that are identified as interesting by Simplified Models and their extensions. This thesis also presented the *Stop Analysis*, a flagship of the searches for Supersymmetry, as a means to constrain Dark Matter production in association with heavy quarks. Started as a simple reinterpretation of the Supersymmetry search, the engagement for Dark Matter constraints in this channel was strengthened and a dedicated optimisation was included within the results presented here. In my opinion, this is an example of where a wider focus of analyses, motivated by similar final states, can be beneficial. Clearly, only by putting together all the different pieces of knowledge, learning more about Dark Matter gets possible. Not one experiment, not one search strategy, not even one discipline of physics alone can hope to pin down its properties, which makes it both challenging and interesting.

Appendix A
Additional Aspects of Dark Matter and Its Properties

A.1 Dark Matter Halo Density Profiles

As discussed in Chap. 3, the question of how the Dark Matter component is distributed in galaxies or clusters is not settled and different proposals are discussed. One of the most commonly assumed density profiles is the pseudo-isothermal halo (e.g. applied by [1]):

$$\rho(r) = \rho_0 \left[1 + \left(\frac{r}{r_c}\right)^2 \right]^{-1}. \tag{A.1}$$

Here, ρ denotes the Dark Matter density as a function of the cluster radius r, where ρ_0 is the finite central density and r_c is the core radius. This model can be at best an approximation, since the estimation of the mass enclosed in a sphere does not converge for infinitely large radii. Guided by numerical simulations of structure formation, Navarro, Frenk and White proposed the following density profile, called NFW [2]:

$$\rho(r) = \frac{\rho_{crit}\delta_c}{\left(\frac{r}{r_s}\right)\left(1+\frac{r}{r_s}\right)^2}, \tag{A.2}$$

where $\rho_{crit} = 3H^2/8\pi G$ is the critical density, δ_c is the so-called characteristic density, a dimension-free parameter, and r_s is the scale radius, denoting the radius at which the slope of the density profile is supposed to change. The NFW profile fits the observations made in very different systems, from single galaxies over galaxy clusters, spanning several orders of magnitude in halo masses, remarkably well while the enclosed mass still diverges. It is common to estimate the mass of the Dark Matter halo by taking the mass that corresponds to a density 200 times larger than the critical density. The agreement with large-scale structure simulations is slightly improved by introducing an additional parameter to the NFW profile, using a so-called Einasto profile [3, 4]:

$$\rho(r) = \rho_e \exp\left[-d_n\left(\left(\frac{r}{r_c}\right)^{\frac{1}{n}} - 1\right)\right]. \tag{A.3}$$

Here, n denotes the additional parameter with d_n being a simple function of n, ensuring that ρ_e is the density at the radius defining a sphere that contains half of the total mass.

A.2 Cosmic Expansion

In principle, an infinite and stable universe is already excluded by the fact that it is dark at night (Olber's Paradox) [5]. Observations revealed that galaxies indeed recede in all directions with velocities proportional to their distance to us. This observation was first made by Hubble in 1929. It is described by the Hubble law:

$$v = H_0 d \tag{A.4}$$

where H_0 denotes the Hubble constant, i.e. the expansion rate of the universe, and has a value of $H_0 = 67.3$ km s^{-1} Mpc^{-1} [6]. Since neither the velocity v nor the distance d of far galaxies can be measured directly, the results are inferred from the luminosity of so-called standard candles, i.e processes whose light emission is always the same and well known, such that the reduction seen in the measured luminosity indicates the distance, and the measured *redshift z*, which leads to the relative velocity of the galaxy:

$$z = \frac{\lambda_{observed} - \lambda_{emitted}}{\lambda_{emitted}} \tag{A.5}$$

The so-called scale factor $a(t)$ can be defined as follows:

$$d = a(t)x, \tag{A.6}$$

where x denotes the present distance, hence today's scale factor is by definition $a_0 = 1$. Clearly, H_0 and a are closely related, namely via:

$$\dot{d} = \dot{a}x, \tag{A.7}$$

$$H(a) \equiv \frac{\dot{a}}{a}. \tag{A.8}$$

The expansion velocity can then be described as:

$$v = \frac{\dot{a}}{a}d = H_0 d. \tag{A.9}$$

Appendix A: Additional Aspects of Dark Matter and Its Properties

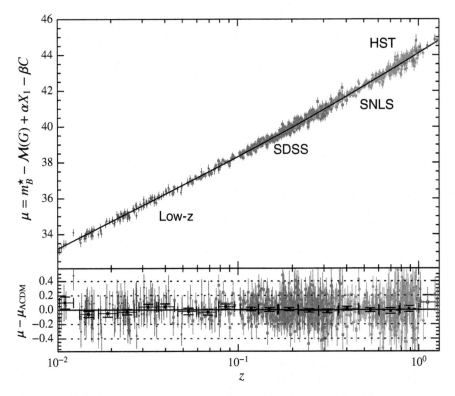

Fig. A.1 Hubble diagram of a combined sample of supernovae measurements from SDSS and SNLS collaborations. The relation of distance to redshift of the best-fit ΛCDM cosmology for a fixed $H_0 = 70 \, \text{km s}^{-1} \, \text{Mpc}^{-1}$ is shown as the black line. Residuals from the best-fit ΛCDM cosmology as a function of redshift are shown in the lower panel. The weighted average of the residuals in logarithmic redshift bins of width $\Delta z/z \sim 0.24$ are shown as black dots. Figure from Ref. [7].

It can also be linked to the above-defined redshift via:

$$a(t_{em}) \equiv (1 + z(t_{em}))^{-1}, \tag{A.10}$$

where t_{em} is the age of the universe at the time the photons were emitted (Fig. A.1).

A.3 Standard Model of Cosmology

The fact that the observed fluctuations in the cosmic microwave background (CMB) are so tiny requires the assumption of a causal connection of regions beyond the event horizon. This presents the strongest hint for the presence of an early inflationary phase of the universe and hence the precise measurement of the CMB and its fluctuations is of great importance to understand better how the universe evolved. This and the assumptions about Dark Matter (DM) described in Chap. 3 shape the so-called *Standard Model of Cosmology*, or ΛCDM where CDM stands for Cold Dark Matter. It is widely successful in describing astrophysical and cosmological observations and it is the simplest model providing explanations of large-scale structures and the distribution of galaxies, the accelerating expansion of the universe, the abundance of hydrogen and other light elements and the anisotropies observed in the CMB. It is based on three major assumptions, motivated by observations: the cosmological principle, which says that the universe is homogeneous and isotropic, the accelerated expansion of metric space and the evolution of the universe from a Big Bang to its present state. The starting-point is given by the Einstein field equation of general relativity:

$$R_{\mu\nu} - \frac{1}{2} g_{\mu\nu} R + \Lambda g_{\mu\nu} = -\frac{8\pi G_N}{c^4} T_{\mu\nu}. \quad (A.11)$$

Resulting from the requirement of invariance under general coordinate transformations and of recovering Newton's law in the limiting case of classical scales, the equation clearly connects the geometry of the universe, in form of its metric $g_{\mu\nu}$ on the left-hand side, with the energy content in form of the energy-momentum tensor $T_{\mu\nu}$. $R_{\mu\nu}$ is the so-called Ricci tensor, R the Ricci scalar, connected to the Riemann curvature. G_N denotes Newton's constant and Λ the cosmological constant. The theory then uses the so-called *FLRW metric* and the *cosmological equation of state* to derive the *Friedmann equations* as a specific solution of the Einstein equations to describe the observable universe from right after the inflationary epoch to present and future.

Metric

The metric assumes the homogeneity and isotropy of space and that the spatial component of the metric can depend on time. The generic metric which meets these conditions and obeys Einstein's field equation was formulated between 1920 and 1930 by Friedmann, Lemaître, Robertson and Walker and is therefore called FLRW metric. It reads:

$$ds^2 = -c^2 dt^2 + a(t)^2 \left(\frac{dr^2}{1 - kr^2} + r^2 d\Omega^2 \right). \quad (A.12)$$

The parameter $a(t)$ is called the *scale factor* and k/a^2 gives the spatial curvature, where k can take the values of $0, \pm 1$. For $k = 0$, the Minkowski metric, describing a flat space, is recovered.

Friedmann Equation

Using the FLRW metric, the components of the Einstein equation leads to the Friedmann equations:

Appendix A: Additional Aspects of Dark Matter and Its Properties

$$\left(\frac{\dot{a}}{a}\right)^2 + \frac{k}{a^2} = \frac{8\pi G_N}{3}\rho_{tot}. \tag{A.13}$$

G_N is Newton's constant, a is again the scale factor and $\dot{a}$ its proper time derivative. ρ_{tot} denotes the total energy density of the universe. One can now again formulate the Hubble parameter as:

$$H(t) = \frac{\dot{a}(t)}{a(t)}. \tag{A.14}$$

Further, it is useful to define the *critical density* as the density for which the universe is flat, i.e. for which $k = 0$, using (A.13):

$$\rho_c = \frac{3H^2}{8\pi G_N}. \tag{A.15}$$

The abundance of a specific component of the universe can then be expressed relative to this critical density:

$$\Omega_i = \frac{\rho_i}{\rho_c}. \tag{A.16}$$

Equation of State

The role of the equation of state (EoS) is to specify the properties of the matter and energy content of the universe. The EoS of a perfect fluid is assumed in the cosmological context:

$$w = p/\rho. \tag{A.17}$$

Here, w is a dimensionless parameter, p denotes the pressure of the system and ρ its energy density. When using the FLRW metric, ρ is related to the scale factor a in the following way:

$$\rho \propto a^{-3(1+w)}. \tag{A.18}$$

The different components of the universe behave differently in terms of their EoS. Ordinary non-relativistic matter, like cold dust, satisfies $p \ll \rho$, leading to $w = 0$. Consequently,

$$\rho_m \propto a^{-3}. \tag{A.19}$$

which can be understood by just considering a volume and the expansion of it. Ultra-relativistic matter, like radiation or matter in the very early universe, behaves like $p = 1/3\rho$, and hence $w = 1/3$. The energy density here decreases more quickly than the volume expansion, because the momentum contribution to the total energy is non-negligible and the wavelength gets red-shifted by the expansion. This results in:

$$\rho_{rad} \propto a^{-4}. \tag{A.20}$$

Dark Energy, in the simplest case of a cosmological constant Λ, obeys $w = -1$, leading to a constant energy density with a described via $a \propto e^{Ht}$. Using these relations, the Friedmann equation can be rewritten as:

$$H(a) = \frac{\dot{a}}{a} = H_0\sqrt{\Omega_m a^{-3} + \Omega_{rad} a^{-4} + \Omega_\Lambda}. \tag{A.21}$$

Measuring the EoS of the universe is an ambitious experimental program, that is hoped to bring new insights into cosmology and the history of the universe. To date, w is measured to be very close to -1, pointing to Dark Energy being related to the cosmological constant and dominating the energy content of the universe.

A second aim of experimental cosmology is to determine the curvature of the universe. Current results have determined that it is almost perfectly flat:

$$\Omega_M m + \Omega_{rad} + \Omega_\Lambda = 1.000 \pm 0.004. \tag{A.22}$$

Therefore, ΛCDM postulates $\Omega = 1$.

A.3.1 Open Questions

Despite the success of the particle DM approach within the ΛCDM model, there remain several unsolved problems, for example the so-called *cusp versus core problem*. It relates to the distribution of DM in galaxies, discussed in Appendix A.1. While galaxy rotation curves strongly suggest a core-like distribution, i.e. a constant DM density close to the centre, cosmological models, incorporated in N-body simulations, require a cusp-like profile, where the DM density increases steeply toward the centre. Possible solutions could come from the inclusion of baryons in the N-body simulation, that probably flatten out the profile close to the centre. Also assuming a small amount of warm DM or DM self interaction could help to solve the problem.

Another open issue is known as the *missing satellites problem*: numerical simulations incorporating the commonly assumed properties of DM predict a lot more dwarf galaxies than actually observed. As an example, one would expect 38 dwarf galaxies in the Milky Way, but only 11 are observed [8]. The missing dwarf galaxies might exist (or have existed) but cannot be observed, either because they are fully DM dominated and did not attract enough baryonic matter to become a luminal galaxy, or because they were stripped by or merged with larger galaxies. Both of these suggestions seem unlikely, which goes under the term of the *too-big-to-fail problem*, stating that a fair fraction of the predicted satellite galaxies should be big and massive enough to be detectable. Also, the size and DM content of the observed satellites does not seem to match the predictions [9].

References

1. J.E. Gunn, J.R. Gott III, On the infall of matter into clusters of galaxies and some effects on their evolution. Astrophys. J. **176**, 1–19 (1972)
2. J.F. Navarro, C.S. Frenk, S.D.M. White, A universal density profile from hierarchical clustering. Astrophys. J. **490**, 493–508 (1997), arXiv:astro-ph/9611107 [astro-ph]
3. J. Einasto, Kinematics and dynamics of stellar systems. Trudy Astrofizicheskogo Instituta Alma-Ata **5**, 87–100 (1965)
4. A.W. Graham, D. Merritt, B. Moore, J. Diemand, B. Terzic, Empirical models for dark matter halos. I. Nonparametric construction of density profiles and comparison with parametric models. Astron. J. **132**, 2685–2700 (2006), arXiv:astro-ph/0509417 [astro-ph]
5. E.R. Harrison, *Darkness at Night: A Riddle of the Universe* (Harvard University Press, Cambridge, MA, 1987)
6. C. Patrignani, Review of particle physics. Chin. Phys. **C40**(10), 100001 (2016)
7. SDSS Collaboration, M. Betoule et al., Improved cosmological constraints from a joint analysis of the SDSS-II and SNLS supernova samples. Astron. Astrophys. **568**, A22 (2014). arXiv:1401.4064 [astro-ph.CO]
8. A.A. Klypin, A.V. Kravtsov, O. Valenzuela, F. Prada, Where are the missing Galactic satellites? Astrophys. J. **522**, 82–92 (1999), arXiv:astro-ph/9901240 [astro-ph]
9. E. Papastergis, R. Giovanelli, M.P. Haynes, F. Shankar, Is there a "too big to fail" problem in the field? Astron. Astrophys. **574**, A113 (2015). arXiv:1407.4665 [astro-ph.GA]

Appendix B
Differential Cross-Sections for Additional Effective Operators

As discussed in Chap. 6, the cross sections of Dark Matter pair production were derived for several effective operators in dependence on the momentum transfer Q_{tr} in order to evaluate the validity of the effective-field-theory approach. Below, the operators not presented in that chapter are listed for completeness.

$$\left.\frac{d^2\hat{\sigma}}{dp_T d\eta}\right|_{D12} = \frac{3\alpha_s^3}{256\pi^2\Lambda^6} \frac{(x_1 x_2 s)^3}{(Q_{tr}^2 - x_1 x_2 s)^2} \frac{Q_{tr}\sqrt{Q_{tr}^2 - 4m_{DM}^2}}{p_T} \left[1 - 4\frac{Q_{tr}^2 - p_T^2}{x_1 x_2 s} + \frac{8Q_{tr}^4 + 21p_T^4}{(x_1 x_2 s)^2} \right.$$
$$- 2Q_{tr}^2 \frac{5Q_{tr}^4 + 4Q_{tr}^2 p_T^2 + 5p_T^4}{(x_1 x_2 s)^3} + Q_{tr}^4 \frac{8Q_{tr}^4 + 8Q_{tr}^2 p_T^2 + 5p_T^4}{(x_1 x_2 s)^4} - 4Q_{tr}^8 \frac{Q_{tr}^2 + p_T^2}{(x_1 x_2 s)^5}$$
$$\left. + \frac{Q_{tr}^{12}}{(x_1 x_2 s)^6} \right], \tag{B.1}$$

$$\left.\frac{d^2\hat{\sigma}}{dp_T d\eta}\right|_{D13} = \frac{3\alpha_s^3}{256\pi^2\Lambda^6} \frac{(x_1 x_2 s)^3}{(Q_{tr}^2 - x_1 x_2 s)^2} \frac{(Q_{tr}^2 - 4m_{DM}^2)^{3/2}}{p_T Q_{tr}} \left[1 - 4\frac{Q_{tr}^2}{x_1 x_2 s} + \frac{8Q_{tr}^4 + 8Q_{tr}^2 p_T^2 + 5p_T^4}{(x_1 x_2 s)^2} \right.$$
$$- 2Q_{tr}^2 \frac{5Q_{tr}^4 + 6Q_{tr}^2 p_T^2 - 3p_T^4}{(x_1 x_2 s)^3} + Q_{tr}^4 \frac{8Q_{tr}^4 + 8Q_{tr}^2 p_T^2 + 5p_T^4}{(x_1 x_2 s)^4} - 4Q_{tr}^8 \frac{Q_{tr}^2 + p_T^2}{(x_1 x_2 s)^5}$$
$$\left. + \frac{Q_{tr}^{12}}{(x_1 x_2 s)^6} \right], \tag{B.2}$$

$$\left.\frac{d^2\hat{\sigma}}{dp_T d\eta}\right|_{D14} = \frac{3\alpha_s^3}{256\pi^2\Lambda^6} \frac{(x_1 x_2 s)^3}{(Q_{tr}^2 - x_1 x_2 s)^2} \frac{Q_{tr}\sqrt{Q_{tr}^2 - 4m_{DM}^2}}{p_T} \left[1 - 4\frac{Q_{tr}^2}{x_1 x_2 s} + \frac{8Q_{tr}^4 + 8Q_{tr}^2 p_T^2 + 5p_T^4}{(x_1 x_2 s)^2} \right.$$
$$- 2Q_{tr}^2 \frac{5Q_{tr}^4 + 6Q_{tr}^2 p_T^2 - 3p_T^4}{(x_1 x_2 s)^3} + Q_{tr}^4 \frac{8Q_{tr}^4 + 8Q_{tr}^2 p_T^2 + 5p_T^4}{(x_1 x_2 s)^4} - 4Q_{tr}^8 \frac{Q_{tr}^2 + p_T^2}{(x_1 x_2 s)^5}$$
$$\left. + \frac{Q_{tr}^{12}}{(x_1 x_2 s)^6} \right]. \tag{B.3}$$

© Springer International Publishing AG, part of Springer Nature 2018
J. Gramling, *Search for Dark Matter with the ATLAS Detector*,
Springer Theses, https://doi.org/10.1007/978-3-319-95016-7

Appendix C
Auxiliary Material for the *Stop* Analysis

C.1 Overview of Additional Signal Regions

While the DM signal regions (SRs) along with their control regions (CRs) and validation regions (VRs) were already defined and discussed in Chap. 9, the other signal regions considered in the analysis are listed here in the following for completeness (Tables C.1 and C.2).

Table C.1 Overview of the event selections for tN SRs and the associated $t\bar{t}$ (TCR), W+jets (WCR), and Wt (STCR) control regions. Round brackets are used to describe lists of values and square brackets denote intervals

Common event selection for tN			
Trigger	E_T^{miss} trigger		
Lepton	Exactly one signal lepton (e, μ), no additional baseline leptons		
Jets	At least two signal jets, and $|\Delta\phi(\text{jet}_i, \vec{p}_T^{\text{miss}})| > 0.4$ for $i \in \{1, 2\}$		
Hadronic τ veto*	Veto events with a hadronic τ decay and $m_{T2}^{\tau} < 80$ GeV		
Variable	SR1	TCR/WCR	STCR
≥ 4 jets with $p_T >$ (GeV)	(80 50 40 40)	(80 50 40 40)	(80 50 40 40)
E_T^{miss} (GeV)	>260	>200	>200
$H_{T,\text{sig}}^{\text{miss}}$	>14	>5	>5
m_T (GeV)	>170	[30, 90]	[30, 120]
am_{T2} (GeV)	>175	[100, 200]/>100	>200
topness	>6.5	>6.5	>6.5
m_{top}^{χ} (GeV)	<270	<270	<270
$\Delta R(b, \ell)$	<3.0	–	–
$\Delta R(b_1, b_2)$	–	–	>1.2
Number of b-tags	≥ 1	$\geq 1/=0$	≥ 2

(continued)

Table C.1 (continued)

Common event selection for tN			
Variable	tN_high	TCR/WCR	STCR
≥4 jets with $p_T >$ (GeV)	(120 80 50 25)	(120 80 50 25)	(120 80 50 25)
E_T^{miss} (GeV)	>450	>300	>250
$E_{T,\perp}^{miss}$ (GeV)	>180	>160	>160
$H_{T,sig}^{miss}$	>22	>15	>10
m_T (GeV)	>210	[30, 90]	[30, 120]
am_{T2} (GeV)	>175	[100, 200]/>100	>200
$\Delta R(b, \ell)$	<2.4	–	–
$\Delta R(b_1, b_2)$	–	–	>1.2
Number of b-tags	≥1	≥1/=0	≥2
Leading large-R jet p_T (GeV)	>290	>290	>290
Leading large-R jet mass (GeV)	>70	>70	>70
$\Delta\phi(\vec{p}_T^{miss}$, 2nd large-R jet)	>0.6	>0.6	>0.6

Table C.2 Overview of the event selections for bC SRs and the associated $t\bar{t}$ (TCR), W+jets (WCR), and Wt (STCR) control regions. Round brackets are used to describe lists of values and square brackets denote intervals. The hadronic tau veto is not applied to the bCbv SR, since the $t\bar{t}$ background is negligible

Common event selection for bC	
Trigger	E_T^{miss} trigger
Lepton	Exactly one signal lepton (e, μ), no additional baseline leptons
Jets	At least two signal jets, and $\|\Delta\phi(\text{jet}_i, \vec{p}_T^{miss})\| > 0.4$ for $i \in \{1, 2\}$
Hadronic τ veto*	Veto events with a hadronic τ decay and $m_{T2}^\tau < 80$ GeV

Variable	bC2x_diag	TCR/WCR	STCR
≥4 jets with $p_T >$ (GeV)	(70 60 55 25)	(70 60 55 25)	(70 60 55 25)
≥2 b-tagged jets with $p_T >$ (GeV)	(25 25)	(25 25)/–	(25 25)
E_T^{miss} (GeV)	>230	>230	>230
$H_{T,sig}^{miss}$	>14	>14	>5
m_T (GeV)	>170	[30, 90]	[30, 120]
am_{T2} (GeV)	>170	[100, 200]/>170	>200
$\|\Delta\phi(\text{jet}_i, \vec{p}_T^{miss})\|$ ($i = 1$)	>1.2	>1.2	>1.2

(continued)

Table C.2 (continued)

Common event selection for bC			
$\|\Delta\phi(\text{jet}_i, \vec{p}_T^{\text{miss}})\|(i=2)$	>0.8	>0.8	>0.8
$\Delta R(b_1, b_2)$	–	–	>1.4
Number of b-tags	≥ 2	$\geq 2/=0$	≥ 2
Variable	bC2x_med	TCR/WCR	STCR
≥ 4 jets with $p_T >$ (GeV)	(170 110 25 25)	(170 110 25 25)	(170 110 25 25)
≥ 2 b-tagged jets with $p_T >$ (GeV)	(105 100)	(105 100)/–	(105 100)
E_T^{miss} (GeV)	>210	>210	>210
$H_{T,\text{sig}}^{\text{miss}}$	>7	>7	>7
m_T (GeV)	>140	[30, 90]	[30, 120]
am_{T2} (GeV)	>210	[100, 210]/>210	>210
$\|\Delta\phi(\text{jet}_i, \vec{p}_T^{\text{miss}})\|(i=1)$	>1.0	>1.0	>1.0
$\|\Delta\phi(\text{jet}_i, \vec{p}_T^{\text{miss}})\|(i=2)$	>0.8	>0.8	>0.8
$\Delta R(b_1, b_2)$	–	–	>1.2
Number of b-tags	≥ 2	$\geq 2/=0$	≥ 2
Variable	bCbv	TCR	WCR
≥ 2 jets with $p_T >$ (GeV)	(120 80)	(120 80)	(120 80)
E_T^{miss} (GeV)	>360	>360	>360
$H_{T,\text{sig}}^{\text{miss}}$	>16	>16	>16
m_T (GeV)	>200	[30, 90]	[30, 90]
Lepton p_T (GeV)	>60	>60	>60
$\|\Delta\phi(\text{jet}_i, \vec{p}_T^{\text{miss}})\|(i=1)$	>2.0	>2.0	>2.0
$\|\Delta\phi(\text{jet}_i, \vec{p}_T^{\text{miss}})\|(i=2)$	>0.8	>0.8	>0.8
Number of b-tags	$=0$	≥ 1	$=0$
Leading large-R jet mass (GeV)	[70, 100]	[70, 100]	[70, 100]
$\Delta\phi(\vec{p}_T^{\text{miss}}, \ell)$	>1.2	–	–

C.2 Full List of Preselection Plots

See Figs. C.1, C.2, C.3, C.4, C.5 and C.6.

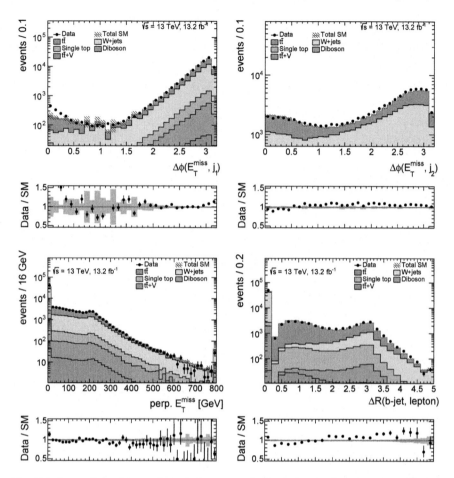

Fig. C.1 Comparison of data and simulation at pre-selection before the fit to data. Only statistical uncertainties are displayed. The last bin includes the overflow. Shown are the angular distance in the transverse plane between E_T^{miss} and the leading (top left) and sub-leading (top right) jet and the lepton (lower left), as well as the radial distance between the b-jet and the lepton (lower right)

Appendix C: Auxiliary Material for the *Stop* Analysis

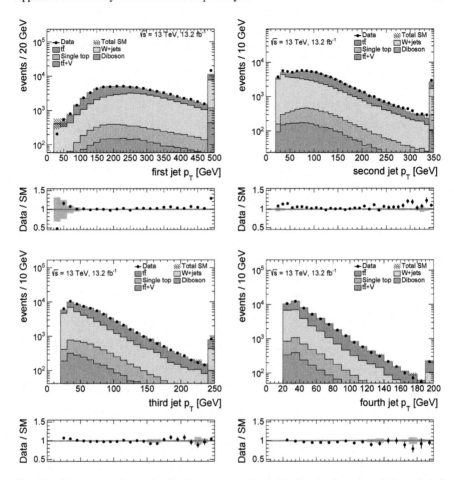

Fig. C.2 Comparison of data and simulation at pre-selection before the fit to data. Only statistical uncertainties are displayed. The last bin includes the overflow. Shown are the distributions of transverse momenta for the four jets in the event

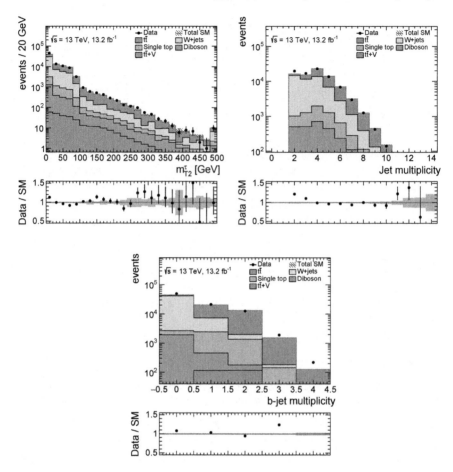

Fig. C.3 Comparison of data and simulation at pre-selection before the fit to data. Only statistical uncertainties are displayed. The last bin includes the overflow. Shown are the perpendicular E_T^{miss} distribution (top left), the variable used in the tau veto (m_{T2}^τ, top right), and the jet (lower left) and b-jet multiplicity (lower right)

Appendix C: Auxiliary Material for the *Stop* Analysis

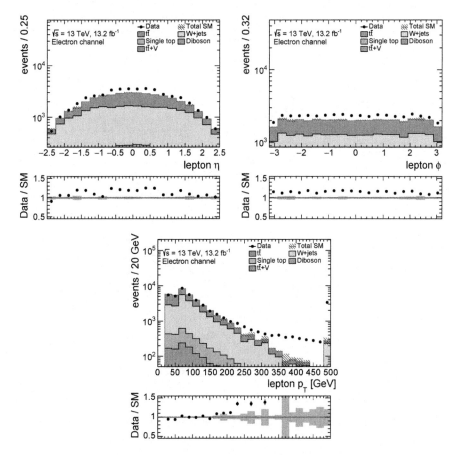

Fig. C.4 Comparison of data and simulation at pre-selection before the fit to data. Only statistical uncertainties are displayed. The last bin includes the overflow. Shown are distributions of lepton pseudo-rapidity (top left), ϕ (top right) and p_T (bottom) in the electron channel

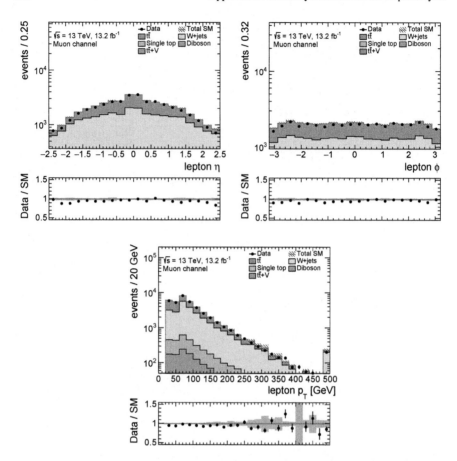

Fig. C.5 Comparison of data and simulation at pre-selection before the fit to data. Only statistical uncertainties are displayed. The last bin includes the overflow. Shown are distributions of lepton pseudo-rapidity (top left), ϕ (top right) and p_T (bottom) in the muon channel

Appendix C: Auxiliary Material for the *Stop* Analysis   237

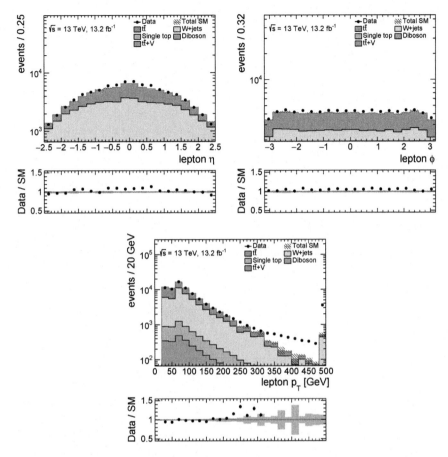

Fig. C.6 Comparison of data and simulation at pre-selection before the fit to data. Only statistical uncertainties are displayed. The last bin includes the overflow. Shown are distributions of lepton pseudo-rapidity (top left), ϕ (top right) and p_T (bottom) (electron and muon channel combined)

C.3 Full List of Control Region Plots

C.3.1 DM_low

See Figs. C.7, C.8, C.9, C.10, C.11, C.12, C.13, C.14, C.15 and C.16.

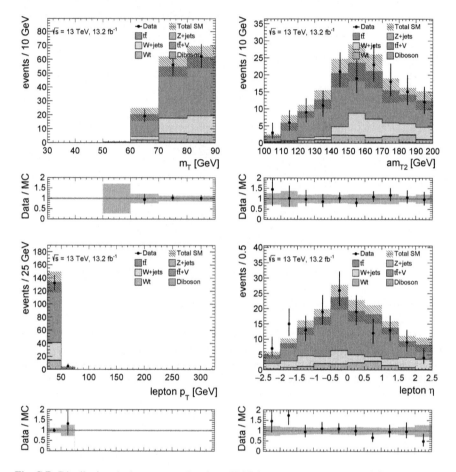

Fig. C.7 Distributions in the top control region of DM_low: transverse mass (top left), asymmetric stransverse mass am_{T2} (top right), lepton p_T and η (bottom left and right). Each background ($t\bar{t}$, W+jets, Wt, and $t\bar{t} + W/Z$) is normalised by normalisation factors obtained by background-only fits. The uncertainty band includes statistical and all experimental systematic uncertainties. The last bin includes overflow

Appendix C: Auxiliary Material for the *Stop* Analysis

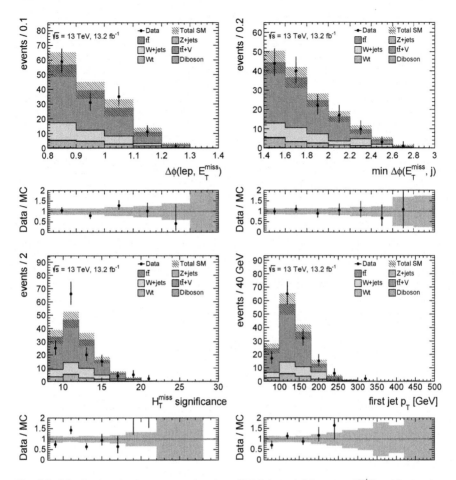

Fig. C.8 Distributions in the top control region of DM_low: $\Delta\phi$ between E_T^{miss} and lepton (top left), minimum $\Delta\Phi$ between E_T^{miss} and any signal jet (top right), $H_{T,sig}^{miss}$ (bottom left) and leading jet p_T (bottom right). Each background ($t\bar{t}$, W+jets, Wt, and $t\bar{t}+W/Z$) is normalised by normalisation factors obtained by background-only fits. The uncertainty band includes statistical and all experimental systematic uncertainties. The last bin includes overflow

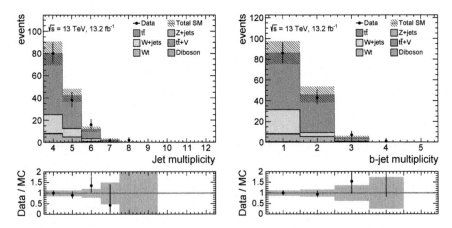

Fig. C.9 Distributions in the top control region of DM_low: jet (left) and b-jet multiplicity (right). Each background ($t\bar{t}$, W+jets, Wt, and $t\bar{t} + W/Z$) is normalised by normalisation factors obtained by background-only fits. The uncertainty band includes statistical and all experimental systematic uncertainties. The last bin includes overflow

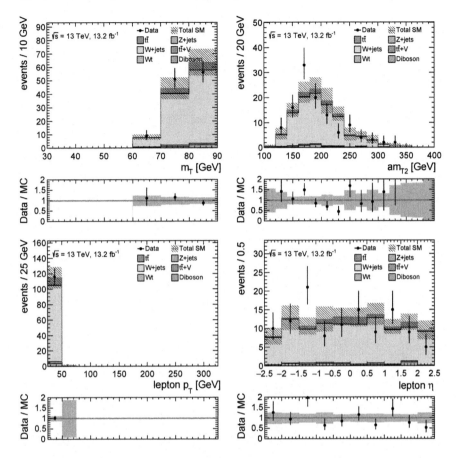

Fig. C.10 Distributions in the W+jets control region of DM_low: transverse mass (top left), asymmetric stransverse mass am_{T2} (top right), lepton p_T and η (bottom left and right). Each background ($t\bar{t}$, W+jets, Wt, and $t\bar{t} + W/Z$) is normalised by normalisation factors obtained by background-only fits. The uncertainty band includes statistical and all experimental systematic uncertainties. The last bin includes overflow

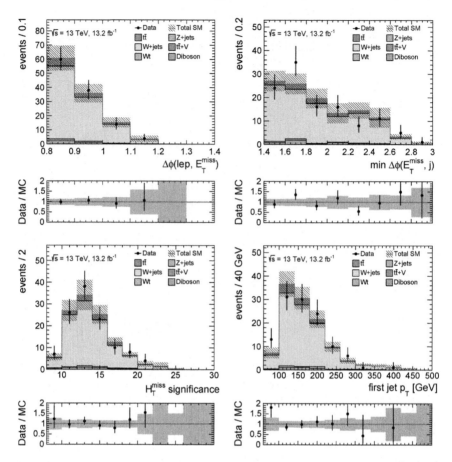

Fig. C.11 Distributions in the W+jets control region of DM_low: $\Delta\phi$ between E_T^{miss} and lepton (top left), minimum $\Delta\Phi$ between E_T^{miss} and any signal jet (top right), $H_{T,sig}^{miss}$ (bottom left) and leading jet p_T (bottom right). Each background ($t\bar{t}$, W+jets, Wt, and $t\bar{t} + W/Z$) is normalised by normalisation factors obtained by background-only fits. The uncertainty band includes statistical and all experimental systematic uncertainties. The last bin includes overflow

Fig. C.12 Distributions in the W+jets control region of DM_low: jet multiplicity. Each background ($t\bar{t}$, W+jets, Wt, and $t\bar{t} + W/Z$) is normalised by normalisation factors obtained by background-only fits. The uncertainty band includes statistical and all experimental systematic uncertainties. The last bin includes overflow

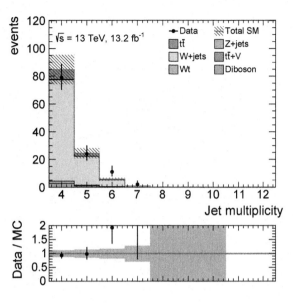

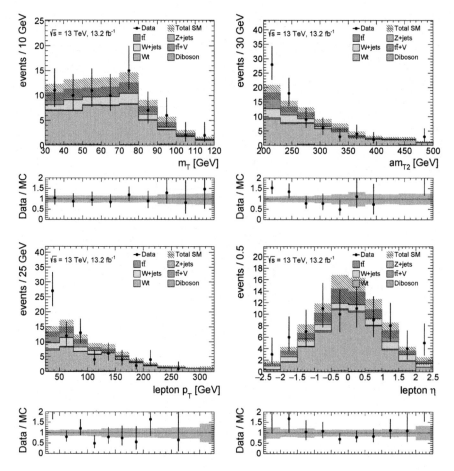

Fig. C.13 Distributions in the single-top control region of DM_low: transverse mass (top left), asymmetric stransverse mass am_{T2} (top right), lepton p_T and η (bottom left and right). Each background ($t\bar{t}$, W+jets, Wt, and $t\bar{t} + W/Z$) is normalised by normalisation factors obtained by background-only fits. The uncertainty band includes statistical and all experimental systematic uncertainties. The last bin includes overflow

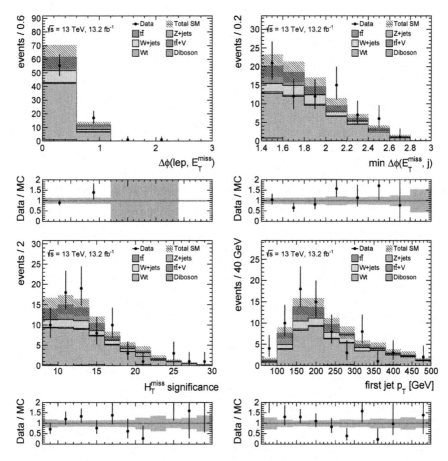

Fig. C.14 Distributions in the single-top control region of DM_low: $\Delta\phi$ between E_T^{miss} and lepton (top left), minimum $\Delta\Phi$ between E_T^{miss} and any signal jet (top right), $H_{T,sig}^{miss}$ (bottom left) and leading jet p_T (bottom right). Each background ($t\bar{t}$, W+jets, Wt, and $t\bar{t} + W/Z$) is normalised by normalisation factors obtained by background-only fits. The uncertainty band includes statistical and all experimental systematic uncertainties. The last bin includes overflow

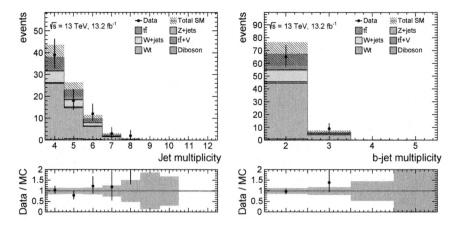

Fig. C.15 Distributions in the single-top control region of DM_low: jet (left) and b-jet multiplicity (right). Each background ($t\bar{t}$, W+jets, Wt, and $t\bar{t} + W/Z$) is normalised by normalisation factors obtained by background-only fits. The uncertainty band includes statistical and all experimental systematic uncertainties. The last bin includes overflow

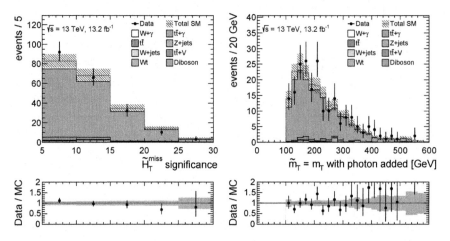

Fig. C.16 Distributions in the $t\bar{t}\gamma$ control region of DM_low: $\Delta\phi$ between E_T^{miss} and lepton (top left), minimum $\Delta\Phi$ between E_T^{miss} and any signal jet (top right), $H_{T,\text{sig}}^{\text{miss}}$ (bottom left) and leading jet p_T (bottom right). Each background ($t\bar{t}$, W+jets, Wt, and $t\bar{t} + W/Z$) is normalised by normalisation factors obtained by background-only fits. The uncertainty band includes statistical and all experimental systematic uncertainties. The last bin includes overflow

Appendix C: Auxiliary Material for the *Stop* Analysis

C.3.2 DM_high

See Figs. C.17, C.18, C.19, C.20, C.21, C.22, C.23, C.24, C.25 and C.26.

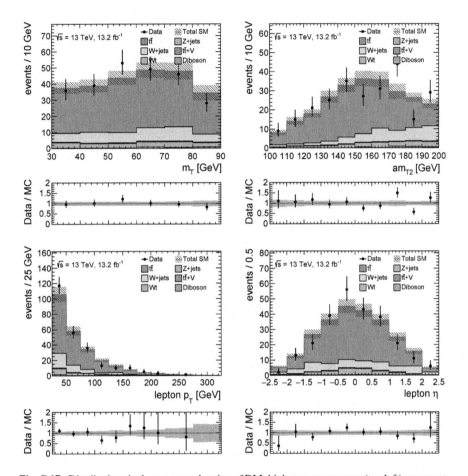

Fig. C.17 Distributions in the top control region of DM_high: transverse mass (top left), asymmetric stransverse mass am_{T2} (top right), lepton p_T and η (bottom left and right). Each background ($t\bar{t}$, W+jets, Wt, and $t\bar{t} + W/Z$) is normalised by normalisation factors obtained by background-only fits. The uncertainty band includes statistical and all experimental systematic uncertainties. The last bin includes overflow

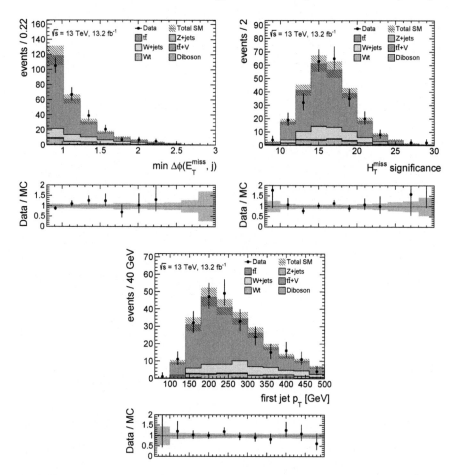

Fig. C.18 Distributions in the top control region of DM_high: minimum $\Delta\Phi$ between E_T^{miss} and any signal jet (top left), $H_{T,\text{sig}}^{\text{miss}}$ (top right) and leading jet p_T (bottom). Each background ($t\bar{t}$, W+jets, Wt, and $t\bar{t}+W/Z$) is normalised by normalisation factors obtained by background-only fits. The uncertainty band includes statistical and all experimental systematic uncertainties. The last bin includes overflow

Appendix C: Auxiliary Material for the *Stop* Analysis

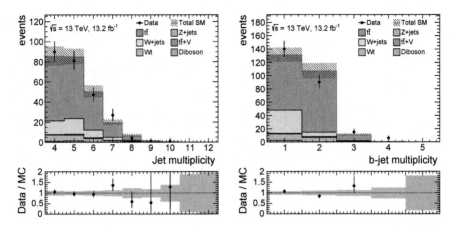

Fig. C.19 Distributions in the top control region of DM_high: jet (left) and b-jet multiplicity (right). Each background ($t\bar{t}$, W+jets, Wt, and $t\bar{t} + W/Z$) is normalised by normalisation factors obtained by background-only fits. The uncertainty band includes statistical and all experimental systematic uncertainties. The last bin includes overflow

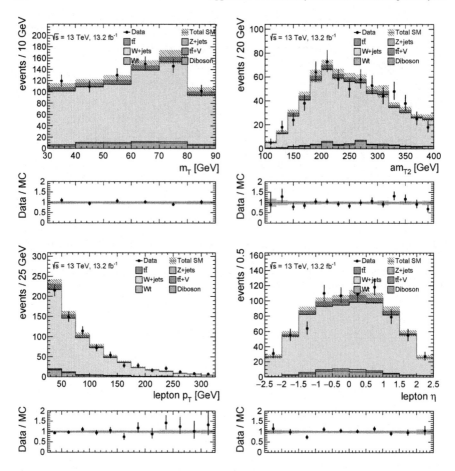

Fig. C.20 Distributions in the W+jets control region of DM_high: transverse mass (top left), asymmetric stransverse mass am_{T2} (top right), lepton p_T and η (bottom left and right). Each background ($t\bar{t}$, W+jets, Wt, and $t\bar{t} + W/Z$) is normalised by normalisation factors obtained by background-only fits. The uncertainty band includes statistical and all experimental systematic uncertainties. The last bin includes overflow

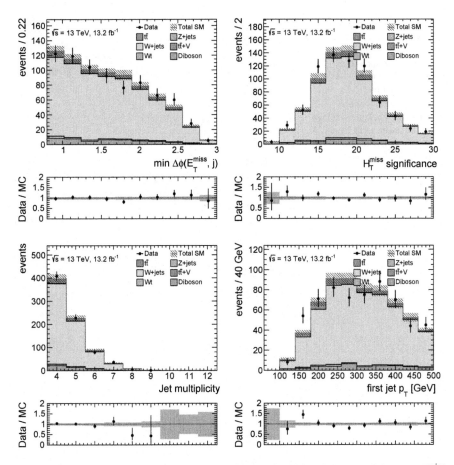

Fig. C.21 Distributions in the W+jets control region of DM_high: minimum $\Delta\Phi$ between E_T^{miss} and any signal jet (top left), $H_{T,sig}^{miss}$ (top right), jet multiplicity (bottom left) and leading jet p_T (bottom right). Each background ($t\bar{t}$, W+jets, Wt, and $t\bar{t} + W/Z$) is normalised by normalisation factors obtained by background-only fits. The uncertainty band includes statistical and all experimental systematic uncertainties. The last bin includes overflow

Fig. C.22 Distributions in the W+jets control region of DM_high: jet multiplicity. Each background ($t\bar{t}$, W+jets, Wt, and $t\bar{t}+W/Z$) is normalised by normalisation factors obtained by background-only fits. The uncertainty band includes statistical and all experimental systematic uncertainties. The last bin includes overflow

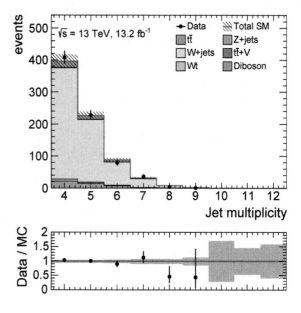

Appendix C: Auxiliary Material for the *Stop* Analysis

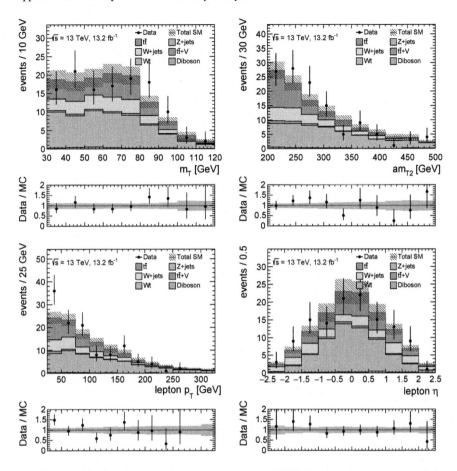

Fig. C.23 Distributions in the single-top control region of DM_high: transverse mass (top left), asymmetric stransverse mass am_{T2} (top right), lepton p_T and η (bottom left and right). Each background ($t\bar{t}$, W+jets, Wt, and $t\bar{t}+W/Z$) is normalised by normalisation factors obtained by background-only fits. The uncertainty band includes statistical and all experimental systematic uncertainties. The last bin includes overflow

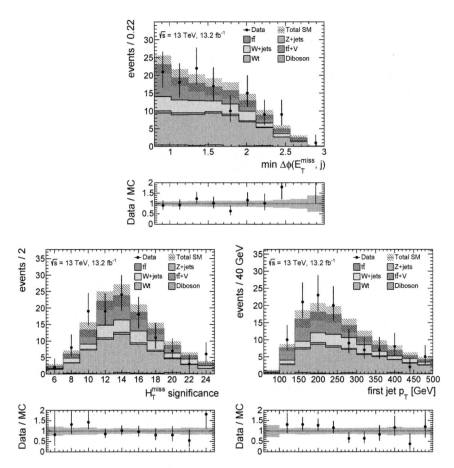

Fig. C.24 Distributions in the single-top control region of DM_high: minimum $\Delta\Phi$ between E_T^{miss} and any signal jet (top right), $H_{T,\mathrm{sig}}^{\mathrm{miss}}$ (bottom left) and leading jet p_T (bottom right). Each background ($t\bar{t}$, W+jets, Wt, and $t\bar{t} + W/Z$) is normalised by normalisation factors obtained by background-only fits. The uncertainty band includes statistical and all experimental systematic uncertainties. The last bin includes overflow

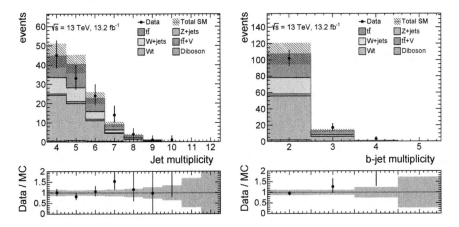

Fig. C.25 Distributions in the single-top control region of DM_high: jet (left) and b-jet multiplicity (right). Each background ($t\bar{t}$, W+jets, Wt, and $t\bar{t} + W/Z$) is normalised by normalisation factors obtained by background-only fits. The uncertainty band includes statistical and all experimental systematic uncertainties. The last bin includes overflow

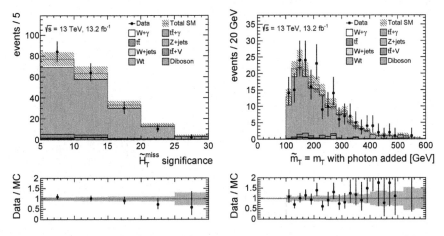

Fig. C.26 Distributions in the $t\bar{t}\gamma$ control region of DM_high: $\Delta\phi$ between E_T^{miss} and lepton (top left), minimum $\Delta\Phi$ between E_T^{miss} and any signal jet (top right), $H_{T,sig}^{miss}$ (bottom left) and leading jet p_T (bottom right). Each background ($t\bar{t}$, W+jets, Wt, and $t\bar{t} + W/Z$) is normalised by normalisation factors obtained by background-only fits. The uncertainty band includes statistical and all experimental systematic uncertainties. The last bin includes overflow

C.4 Full List of Validation Region Plots

C.4.1 DM_low

See Figs. C.27, C.28, C.29, C.30, C.31 and C.32.

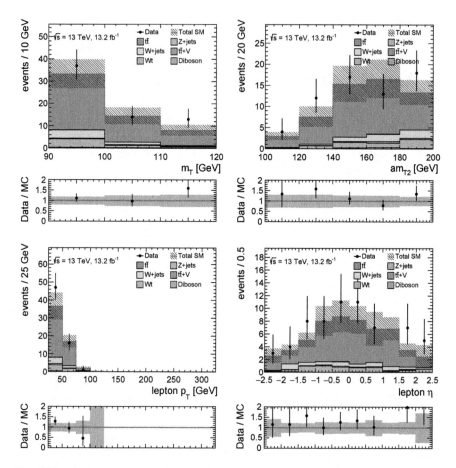

Fig. C.27 Distributions in the top validation region of DM_low: transverse mass (top left), asymmetric stransverse mass am_{T2} (top right), lepton p_T and η (bottom left and right). Each background ($t\bar{t}$, W+jets, Wt, and $t\bar{t} + W/Z$) is normalised by normalisation factors obtained by background-only fits. The uncertainty band includes statistical and all experimental systematic uncertainties. The last bin includes overflow

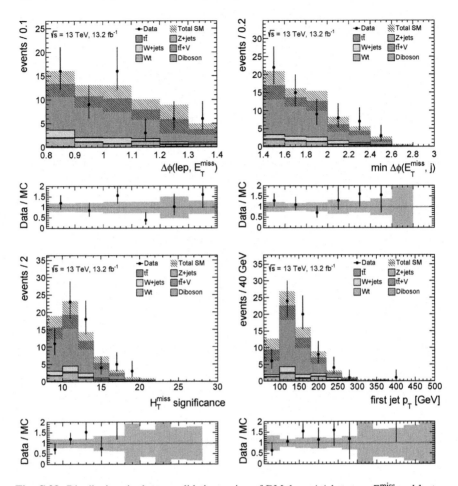

Fig. C.28 Distributions in the top validation region of DM_low: $\Delta\phi$ between E_T^{miss} and lepton (top left), minimum $\Delta\Phi$ between E_T^{miss} and any signal jet (top right), $H_{T,\text{sig}}^{\text{miss}}$ (bottom left) and leading jet p_T (bottom right). Each background ($t\bar{t}$, W+jets, Wt, and $t\bar{t}+W/Z$) is normalised by normalisation factors obtained by background-only fits. The uncertainty band includes statistical and all experimental systematic uncertainties. The last bin includes overflow

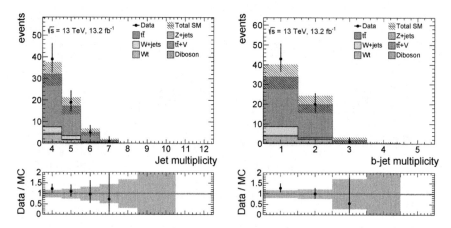

Fig. C.29 Distributions in the top validation region of DM_low: jet (left) and b-jet multiplicity (right). Each background ($t\bar{t}$, W+jets, Wt, and $t\bar{t} + W/Z$) is normalised by normalisation factors obtained by background-only fits. The uncertainty band includes statistical and all experimental systematic uncertainties. The last bin includes overflow

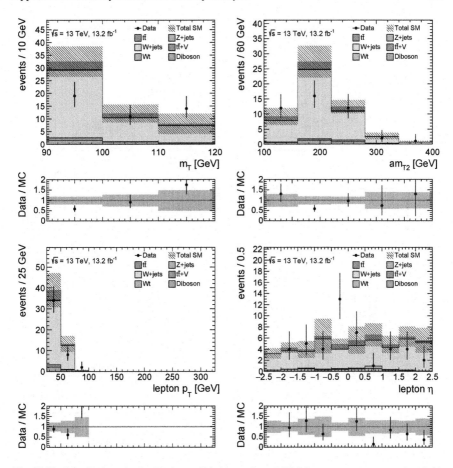

Fig. C.30 Distributions in the W+jets validation region of DM_low: transverse mass (top left), asymmetric stransverse mass am_{T2} (top right), lepton p_T and η (bottom left and right). Each background ($t\bar{t}$, W+jets, Wt, and $t\bar{t}+W/Z$) is normalised by normalisation factors obtained by background-only fits. The uncertainty band includes statistical and all experimental systematic uncertainties. The last bin includes overflow

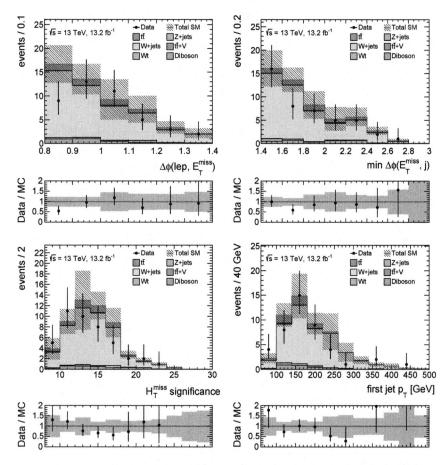

Fig. C.31 Distributions in the W+jets validation region of DM_low: $\Delta\phi$ between E_T^{miss} and lepton (top left), minimum $\Delta\Phi$ between E_T^{miss} and any signal jet (top right), $H_{T,sig}^{miss}$ (bottom left) and leading jet p_T (bottom right). Each background ($t\bar{t}$, W+jets, Wt, and $t\bar{t}+W/Z$) is normalised by normalisation factors obtained by background-only fits. The uncertainty band includes statistical and all experimental systematic uncertainties. The last bin includes overflow

Fig. C.32 Distributions in the W+jets validation region of DM_low: jet multiplicity. Each background ($t\bar{t}$, W+jets, Wt, and $t\bar{t} + W/Z$) is normalised by normalisation factors obtained by background-only fits. The uncertainty band includes statistical and all experimental systematic uncertainties. The last bin includes overflow

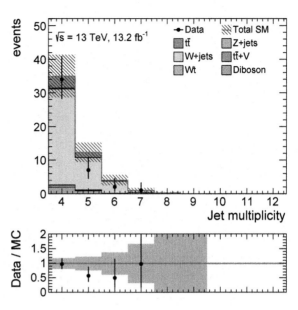

C.4.2 DM_high

See Figs. C.33, C.34, C.35, C.36, C.37 and C.38.

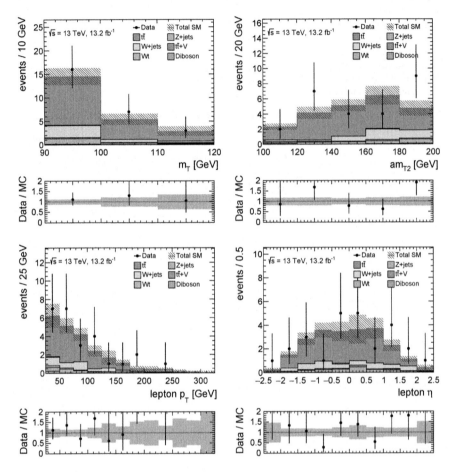

Fig. C.33 Distributions in the top validation region of DM_high: transverse mass (top left), asymmetric stransverse mass am_{T2} (top right), lepton p_T and η (bottom left and right). Each background ($t\bar{t}$, W+jets, Wt, and $t\bar{t} + W/Z$) is normalised by normalisation factors obtained by background-only fits. The uncertainty band includes statistical and all experimental systematic uncertainties. The last bin includes overflow

Appendix C: Auxiliary Material for the *Stop* Analysis

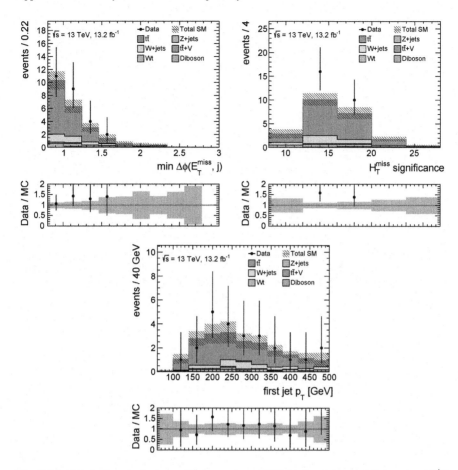

Fig. C.34 Distributions in the top validation region of DM_high: minimum $\Delta\Phi$ between E_T^{miss} and any signal jet (top left), $H_{T,\text{sig}}^{\text{miss}}$ (top right) and leading jet p_T (bottom). Each background ($t\bar{t}$, W+jets, Wt, and $t\bar{t} + W/Z$) is normalised by normalisation factors obtained by background-only fits. The uncertainty band includes statistical and all experimental systematic uncertainties. The last bin includes overflow

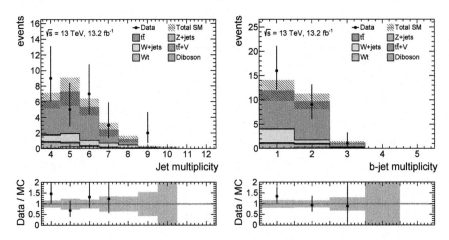

Fig. C.35 Distributions in the top validation region of DM_high: jet (left) and b-jet multiplicity (right). Each background ($t\bar{t}$, W+jets, Wt, and $t\bar{t} + W/Z$) is normalised by normalisation factors obtained by background-only fits. The uncertainty band includes statistical and all experimental systematic uncertainties. The last bin includes overflow

Appendix C: Auxiliary Material for the *Stop* Analysis

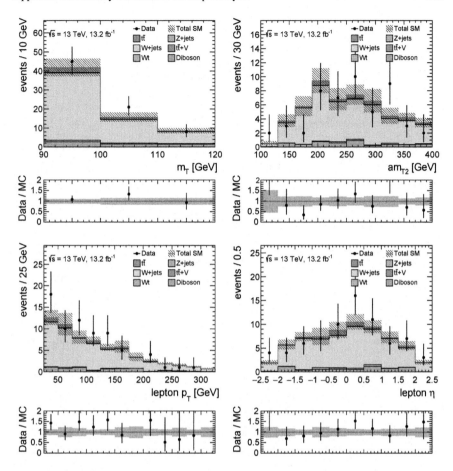

Fig. C.36 Distributions in the W+jets validation region of DM_high: transverse mass (top left), asymmetric stransverse mass am_{T2} (top right), lepton p_T and η (bottom left and right). Each background ($t\bar{t}$, W+jets, Wt, and $t\bar{t} + W/Z$) is normalised by normalisation factors obtained by background-only fits. The uncertainty band includes statistical and all experimental systematic uncertainties. The last bin includes overflow

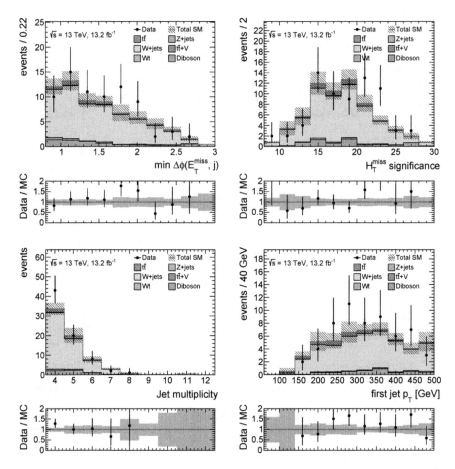

Fig. C.37 Distributions in the W+jets validation region of DM_high: minimum $\Delta\Phi$ between E_T^{miss} and any signal jet (top left), $H_{T,sig}^{miss}$ (top right), jet multiplicity (bottom left) and leading jet p_T (bottom right). Each background ($t\bar{t}$, W+jets, Wt, and $t\bar{t} + W/Z$) is normalised by normalisation factors obtained by background-only fits. The uncertainty band includes statistical and all experimental systematic uncertainties. The last bin includes overflow

Fig. C.38 Distributions in the W+jets validation region of DM_high: jet multiplicity. Each background ($t\bar{t}$, W+jets, Wt, and $t\bar{t} + W/Z$) is normalised by normalisation factors obtained by background-only fits. The uncertainty band includes statistical and all experimental systematic uncertainties. The last bin includes overflow

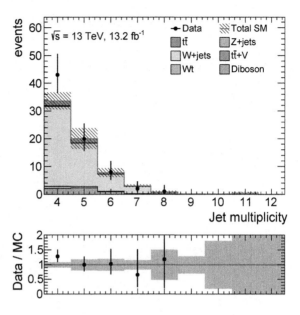

C.5 Full List of Signal Region Plots

C.5.1 DM_low

See Figs. C.39, C.40 and C.41.

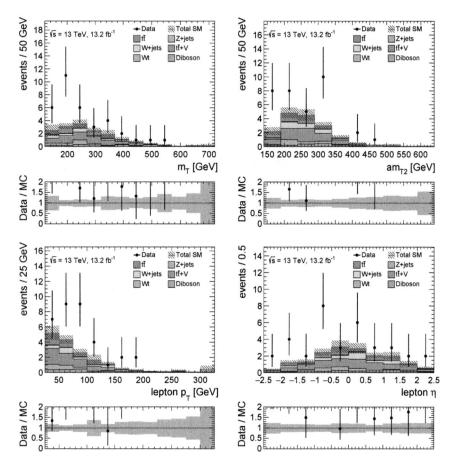

Fig. C.39 Distributions in the DM_low signal region: transverse mass (top left), asymmetric stransverse mass am_{T2} (top right), lepton p_T and η (bottom left and right). Each background ($t\bar{t}$, W+jets, Wt, and $t\bar{t} + W/Z$) is normalised by normalisation factors obtained by background-only fits. The uncertainty band includes statistical and all experimental systematic uncertainties. The last bin includes overflow

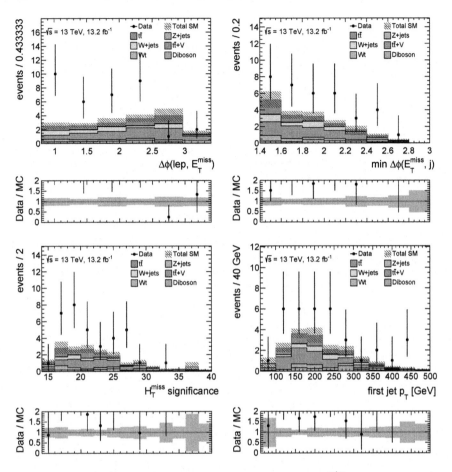

Fig. C.40 Distributions in the DM_low signal region: $\Delta\phi$ between E_T^{miss} and lepton (top left), minimum $\Delta\Phi$ between E_T^{miss} and any signal jet (top right), $H_{T,sig}^{miss}$ (bottom left) and leading jet p_T (bottom right). Each background ($t\bar{t}$, W+jets, Wt, and $t\bar{t}+W/Z$) is normalised by normalisation factors obtained by background-only fits. The uncertainty band includes statistical and all experimental systematic uncertainties. The last bin includes overflow

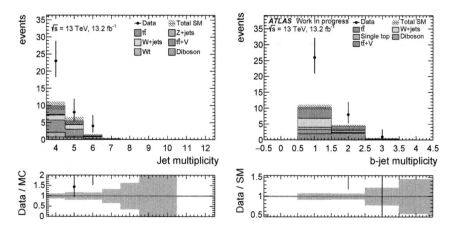

Fig. C.41 Distributions in the DM_low signal region: jet (left) and b-jet multiplicity (right). Each background ($t\bar{t}$, W+jets, Wt, and $t\bar{t} + W/Z$) is normalised by normalisation factors obtained by background-only fits. The uncertainty band includes statistical and all experimental systematic uncertainties. The last bin includes overflow

C.5.2 DM_high

See Figs. C.42, C.43 and C.44.

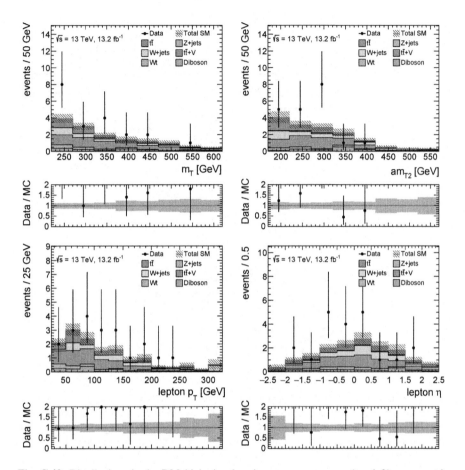

Fig. C.42 Distributions in the DM_high signal region: transverse mass (top left), asymmetric stransverse mass am_{T2} (top right), lepton p_T and η (bottom left and right). Each background ($t\bar{t}$, W+jets, Wt, and $t\bar{t} + W/Z$) is normalised by normalisation factors obtained by background-only fits. The uncertainty band includes statistical and all experimental systematic uncertainties. The last bin includes overflow

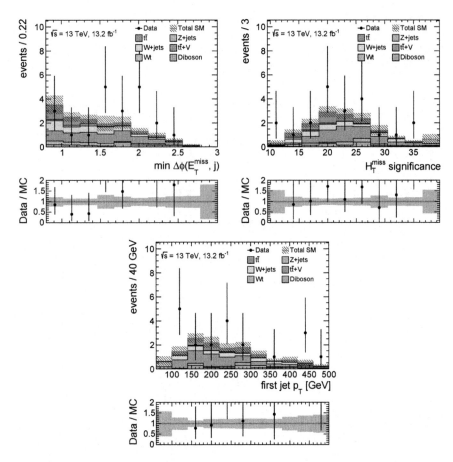

Fig. C.43 Distributions in the DM_high signal region: minimum $\Delta\Phi$ between E_T^{miss} and any signal jet (top left), $H_{T,sig}^{miss}$ (top right) and leading jet p_T (bottom). Each background ($t\bar{t}$, W+jets, Wt, and $t\bar{t} + W/Z$) is normalised by normalisation factors obtained by background-only fits. The uncertainty band includes statistical and all experimental systematic uncertainties. The last bin includes overflow

Fig. C.44 Distributions in the DM_high signal region: jet multiplicity. Each background ($t\bar{t}$, W+jets, Wt, and $t\bar{t} + W/Z$) is normalised by normalisation factors obtained by background-only fits. The uncertainty band includes statistical and all experimental systematic uncertainties. The last bin includes overflow

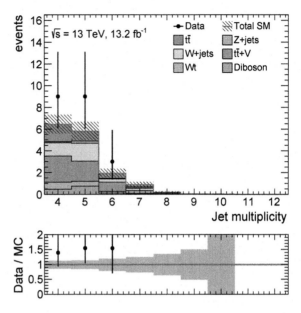

Appendix D
Investigation of the Data Excess

With the observation of an excess over 3σ many additional cross-checks were performed to scrutinise the analysis procedure and its results. A subset is presented in the following, focussing on the DM_low signal region.

D.1 Characteristics of the Excess

As a first approach to understand whether an experimental problem led to the observed excess of data events over the Standard Model background prediction, general properties of the selected events and the excess are reviewed. Neither the event displays (an example is shown in Fig. D.1) nor the detailed listing of signal event properties revealed anything suspicious. In the event display, it can be clearly seen that an event topology in which all reconstructed objects are close by, recoiling against E_T^{miss} is selected by the signal region requirements.

As discussed in Chap. 9, the excess tends to be located towards low m_T, low E_T^{miss} and probably lower jet multiplicities in all three signal regions that observed deviations from the prediction.

Furthermore, small distances between the jets are preferred, in the transverse plane and in terms of ΔR. Exemplarily, the distance between leading and sub-leading or third-leading jet in the transverse plane is shown in Fig. D.2 in the DM_low signal region and its $t\bar{t}$ control region.

Also the overlap between the signal regions that see an excess was mentioned in Chap. 9. This was studied as well, first by quantifying the fraction of data events that appear in more than one signal region. For example, DM_low shares 30% of its events with SR1 and 20% with bC2x_diag. For backgrounds, a similar conclusion holds for more signal-like processes such as $t\bar{t}Z$, while in $t\bar{t}$ the selected events differ between the signal regions. Another way to examine the overlap was to define exclusive signal regions. While in the exclusive DM_low and bC2x_diag the excess is still observed, the exclusive SR1 yield agrees with predictions.

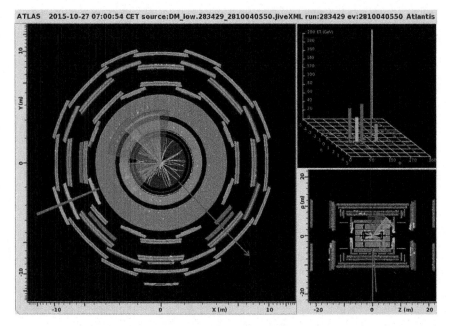

Fig. D.1 Graphical display of Event 2810040550 of Run 283429, passing the DM_low selection. Note that analysis-level objects might differ from those displayed here. The green arrow denotes a muon, the coloured cones represent jets and the red arrow indicates the E_T^{miss} in the event

D.2 Scrutinising the Background Estimate

No apparent mismodelling could be observed in the control and validation regions. As an additional cross-check, the modelling of the top system was checked in the p_T of the reconstructed hadronic top and the $t\bar{t}$ system (Fig. D.3). No significant mismodelling in the control region is observed and no tendency of the excess with respect to these variables is visible.

The possibility of a background that is not considered entering the selection was investigated as well. While for previous iterations of the analysis the Z+jets background was found to be negligible, it was observed to be small but worth considering ($\mathcal{O}(1\%)$) and hence got re-introduced in the analysis chain. Other small backgrounds were checked and confirmed to be negligible ($tttt, tZ, ttWW, tZW$). Backgrounds containing Higgs bosons were not explicitly checked since the relevant data format was not available at that time, however they are expected to be negligible due to the strict requirement on E_T^{miss} that is applied in the analysis.

Appendix D: Investigation of the Data Excess

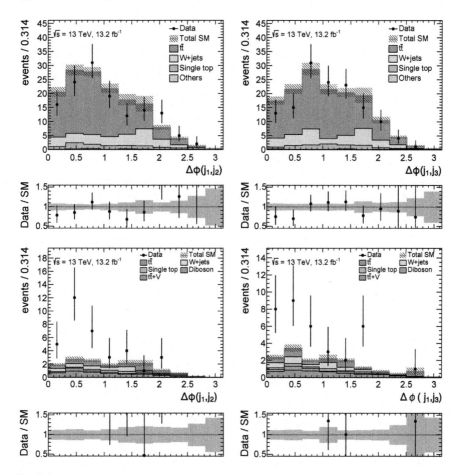

Fig. D.2 Distribution of $\Delta\phi$ between leading and sub-leading (left) or third-leading jet (right) in the top control region (top) and the DM_low signal region (bottom)

Remarkably, a normalisation factor of roughly 1.5 was observed for the $t\bar{t}Z$ background. This could be due to a missing background in the $t\bar{t}\gamma$ control region. Indeed, $W + \gamma$ processes account for about 10% of the events in the $t\bar{t}\gamma$ control region. Including this background in the determination of the normalisation factor led to a slightly smaller normalisation factor for $t\bar{t}Z$. The high normalisation factor was further scrutinised in a tri-lepton $t\bar{t}Z$ validation region. Due to limited statistics such a region is not straight-forward to consider. The discrepancy between data and MC prediction observed in this region was consistent with the normalisation factor.

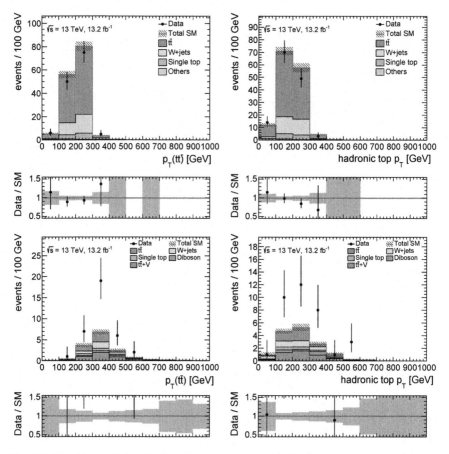

Fig. D.3 Distribution of $t\bar{t}$ p_T (left) and hadronic top p_T (right) in the top control region (top) and the DM_low signal region (bottom)

D.3 Object Reconstruction

Since the analysis relies heavily on E_T^{miss}, special emphasis was given to the validation of its performance. Figure D.4 shows the distribution of the different components of E_T^{miss} in the top control region, Fig. D.5 in the DM_low signal region.

It was also verified that no regions in the η–ϕ plane are present in which unusually many objects are reconstructed (*hot spots*). This could occur in case of an unidentified detector problem. Neither for electrons, nor for muons or jets such a problem was found.

Appendix D: Investigation of the Data Excess

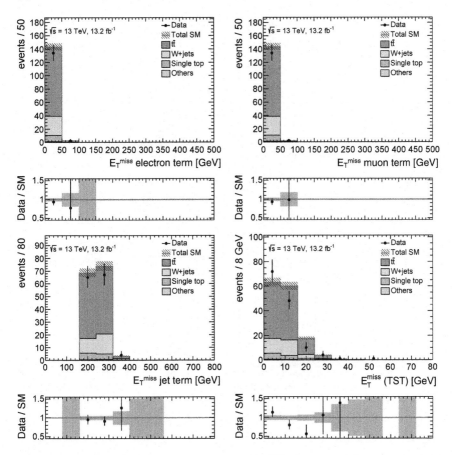

Fig. D.4 Distribution of the different components of the E_T^{miss}: electron (top left), muon (top right), jet (bottom left) and soft term (bottom right) in the top control region of the DM_low signal region (bottom)

Furthermore it was confirmed that there is no significant difference observed between electron and muon channel. In order to exclude any problem stemming from low-quality leptons or badly modelled lepton isolation, different isolation working points corresponding to different isolation criteria were monitored. As shown in Fig. D.6, most electrons and muons fulfil even the strictest requirements. Also the possibility for a jet to originate from pile-up was found to tend to the maximum value, i.e. to the best reconstruction quality for all four signal jets.

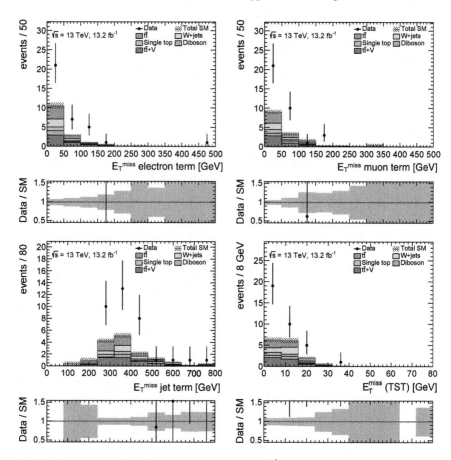

Fig. D.5 Distribution of the different components of the E_T^{miss}: electron (top left), muon (top right), jet (bottom left) and soft term (bottom right) in the DM_low signal region (bottom)

The excess of DM_low prefers exactly one b-jet in the event. This is not the case for SR1 and bC2x_diag. However, it is still important to verify that badly reconstructed b-jets do not cause the observed disagreement between data and MC. As for the leptons, different b-tagging working points were used to study the behaviour of the b-tagging discriminant, since this was not available in the reduced analysis data format. Figure D.7 shows the fulfilled working point requirements exemplarily for the two leading jets. If the jet is tagged as a b-jet it often fulfilled the strictest criteria. Furthermore, the excess does not favour low-quality b-jets.

Appendix D: Investigation of the Data Excess

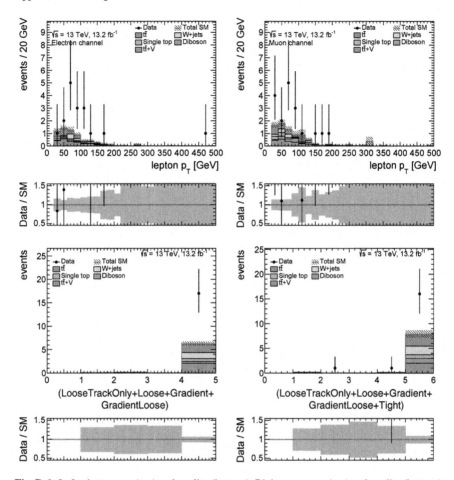

Fig. D.6 Left: electron p_T (top) and quality (bottom). Right: muon p_T (top) and quality (bottom). All are shown for the DM_low signal region

D.4 Conditions of Data Taking

The event yield at preselection and in the signal regions was monitored over different runs and data taking periods and found to be stable. Furthermore, the data sample was split up into a low-pile-up ($\mu < 15$) and a high-pile-up ($\mu > 15$) sample. No significant differences were observed. Exemplarily, Fig. D.8 shows the E_T^{miss} distribution under both conditions.

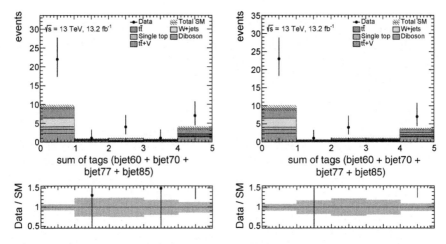

Fig. D.7 Different b-tagging working points tested for the leading (left) and the sub-leading jet (right). All are shown for the DM_low signal region

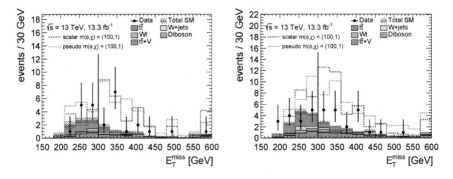

Fig. D.8 Distribution of E_T^{miss} in the DM_low signal region for $\mu < 15$ (left) and $\mu > 15$

Printed by Printforce, the Netherlands